TELEVISION AND ITS AUDIENCE

Television and its Audience

INTERNATIONAL RESEARCH PERSPECTIVES

*A Selection of Papers from the Second
International Television Studies Conference
London, 1986*

EDITED BY PHILLIP DRUMMOND AND RICHARD PATERSON

BFI Publishing

British Film Institute
1988

First published in 1988 by the
British Film Institute
21 Stephen Street
London W1P 1PL

Cover design: Lee Robinson

Set in Century Schoolbook by
Fakenham Photosetting Limited
Fakenham, Norfolk
Printed and bound in Great Britain by
Anchor Brendon Ltd, Tiptree, Essex

British Library Cataloguing in Publication Data

Television and its audience: international
   research perspectives
   1. Television programmes. Audiences.
   I. Drummond, Phillip, *1948–*   II. Paterson,
   Richard, *1947–*   III. British Film
   Institute
   302.2'345

ISBN 0–85170–224–4

Contents

Foreword

The second International Television Studies Conference, held in London from 10 to 12 July 1986, brought together some 450 delegates to receive and debate more than 120 papers from 33 countries. ITSC is a unique venture, co-organised by the British Film Institute and the University of London Institute of Education, with valuable support for the 1986 event, in cash and in kind, from a variety of organisations including the Economic and Social Research Council, Channel Four Television, BBC Enterprises and the British Council.

Conference papers covered six major themes: the media and political economy, with special reference to new developments; the media and politics; media and education; the history of the media; analysis and interpretation of media content; and audience research. The television viewer once more occupied the centre of the stage, and the conference brought home the extent to which the television industry is intimately related to the political, social and cultural characteristics of the society. The diversity of Television Studies was reflected both in the papers presented and in the lively discussions that followed. The conference was marked by a high standard of debate, offered exciting new directions of enquiry and provided a much needed forum for new and old colleagues from across the world to forge or to renew contacts.

The conference had all the hallmarks of success: a diversity of good quality papers, discussants who took their role seriously, the coming together of a galaxy of experts from all parts of the world and an exceptionally well planned and agreeable atmosphere in which both friendship and professional exchange could flourish.

If the advantages of an international conference as successful as this are very great, so are the labours involved in organising it. In this respect, Richard Paterson and Phillip Drummond, together with a small and devoted staff, did a tremendous job, all the more remarkable as they organised this conference only two years after an earlier and equally successful conference of which they were also the architects.

HILDE HIMMELWEIT
Emeritus Professor of Social Psychology,
London School of Economics

Acknowledgments

We are very grateful to the institutions and individuals who gave of their time, energy, and material support to make ITSC 2 and the current volume possible.

We would first like to thank our host institutions, the British Film Institute's TV Unit, and the University of London Institute of Education's Department of English and Media Studies, for their consistent and imaginative long-term support. Our partial independence has been further fortified by generous financial support from other agencies within the culture. Our major sponsor in this category has been the Economic and Social Research Council, whose extremely generous financial aid enabled the event to grow to a truly international size. The British Council were again kindly able to assist with the expenses incurred in connection with overseas visitors to the Conference, while we are grateful to the Canadian High Commission for specific help in respect of visitors from Canada.

Other sponsors were able to offer help in kind. Much of the social dynamic of the event was propelled by the speakers' receptions hosted by Channel Four and by the University of London Institute of Education, whilst all delegates were grateful for the major festivity offered by BBC Enterprises at the Hotel Russell. We are also grateful to Atlas Leisure for facilitating the satellite feed of Music Box and Cable News Network which came to us courtesy of the Music Channel and Turner Broadcasting International, and to Westminster Cable for inviting delegates aboard their travelling cable Starship.

Where individual thanks are due, our first debt is to our Honorary Presidents, Professor Hilde Himmelweit and the late Professor Raymond Williams. We are particularly grateful for Professor Himmelweit's powerful and influential support throughout the period of preparation. We are also grateful to the members of our broadly-based Organising Committee for all their work and endeavour over many months: John Akomfrah, Karen Alexander, Charlotte Brunsdon, David Buckingham, Lesley Caldwell, Richard Collins, Jim Cook, Bob Ferguson, Joy Leman, Caroline Merz, Jim Pines, and Mary Wood. We would also like to thank the numerous postgraduate media students of the University of London Institute of Education, brilliantly marshalled by Bernard Anderson, for the energy and dedication which they brought to their role as Conference stewards, making no small contribution to the smooth running of a large and complicated event.

Two final categories of person deserve our gratitude. The first group has a membership of only three. They are our Conference administrators Lucy Douch, Lilie Ferrari (both BFI) and Susan Gibbons (Institute of Education), who carried huge organisational burdens, far from the public eye, for many months. Not the least of their tasks was the organisation not simply of the Conference itself, but frequently of its Directors, too. The second of these groups is much bigger; it is nothing less than our audience of speakers, chairpersons, discussants, and delegates, a total of some 450 people from over 30 countries. We are grateful to them all for their support, and hope that the event repaid its debt by proving a success in individual terms as well as marking an important stage of development in the institutional field of Television Studies across the globe.

Our final thanks go to Geoffrey Nowell-Smith and Roma Gibson, of BFI Publishing, for being willing, yet again, to become involved in the gruelling task of turning a conference into a book.

PHILLIP DRUMMOND
RICHARD PATERSON
Co-Directors, ITSC 1986

PHILLIP DRUMMOND AND RICHARD PATERSON

Introducing the Television Audience

Study of the audience is central to any understanding of the role of television in contemporary society. Whenever the relationship of television to other institutions – government, the economy, the family, the advertisers – is in question, the nature of the television audience is in question, too. Who views, and why? What, and how, do they watch? What does television viewing do for them – and, in turn, for television itself? In the era of continuing technological and economic change – the era of the video-recorder, empowering the audience as programme-planner; the era of the satellite and increasing deregulation – those questions are still more potent and more urgent.

The audience is certainly an economic category, a collectivity of consumers who participate, through their choice of TV programmes – and hence TV *companies* – in the busily evolving marketplace of commercial mass culture. Viewers are moreover psychological subjects whose perceptions are both developed and constrained by the particular representational regimes of the television discourse and the conditions of TV spectatorship. The television audience is therefore simultaneously an ideological subject whose economic relationship to consumption and whose psychological relationship to perception must be related to other social and political forces. Here the familial and national character of the television audience are crucial issues for analysis.

Various perspectives on these issues are proposed in this selection as we bring together a range of international experiences, and a broad spectrum of disciplinary expertise. The eight groups of essays in this volume – a mere fraction of the research on offer at the Second International Television Studies Conference – suggest sometimes complementary, sometimes very differing approaches to the economic, social, psychological and ideological complexities of studying the television audience.

Modalities
The first category of research – 'Modalities' – is concerned with the relationship between the forms of television and the lived experience of the television audience. Paddy Scannell explores some of the ways

1

in which broadcasting sustains the lives and routines of whole populations, while Claus-Dieter Rath examines the fabled 'liveness' of the television product as a key ontological effect for TV audiences. More specifically, Mary Beth Haralovich studies the historically defined connections between US television sitcoms and the post-war economic boom.

Scannell, in 'Radio Times: The Temporal Arrangements of Broadcasting in the Modern World', reviews the notion of a broadcast calendar and describes its development as part of the national culture in the UK from the earliest days of radio. He contrasts this notion of time in relation to broadcast forms with the specificities of the continuous serials. Serials Scannell sees as narratives which run in parallel with and interact with the actualities of the lived world.

The central conclusion of the essay is that the fundamental function of national broadcasting systems is to mediate modernity by normalising the public sphere and socialising the private. This important conclusion overcomes the oversimple binary opposition between private and public spheres, and conceptualises them as routinely implicated with each other. Clearly any analysis of the social and cultural force of television has to come to terms with this regulatory and normative impact of the medium.

Of interest to Rath, in his essay 'Live/Life: Television as a Generator of Events in Everyday Life', is the way in which citizens in the modern world are undergoing a permanent initiation in which television places events in a fictitious historical stream, filling everyday life with public and publicised events. His review of the live television mode with its inherent high risks, which are accepted because liveness is seen to be a 'new' quality to modern television, leads him to conclude that television's main role is to allow spectators to be 'merely there'. Television events rebind the viewer into the symbolic order so that, for instance, *Live Aid* establishes the spectator as part of an imaginary totality.

Haralovich attends to the historical specificities of television in the home of the 50s in the USA and to the textual systems of two important sitcoms of the period. Her essay, 'Suburban Family Sitcoms and Consumer Product Design: Addressing the Social Subjectivity of Homemakers in the 1950s', reviews the historical determinations of gender representations by examining the representational systems of two key sitcoms – *Father Knows Best* and *Leave it to Beaver*. She situates them within a set of social practices which assisted in the reconstitution and socialisation of the US family in the post-World War Two period.

Suburban development in the 1950s, Haralovich argues, proceeded according to precise ideological considerations concerning the 'zoning' of gender, race, age and social class. The nuclear family was

2

privileged as the central unit of home-ownership and neighbourhood development, and within the home domestic architecture was designed 'to display class attributes and reinforce gender-specific functions of domestic space'. The home-maker's life in the home became, in turn, the focus for the consumer product industry, with its technologically enabled promise of aesthetically enhanced leisure from household drudgery.

The suburban TV family sitcom, in re-articulating many of these concerns, thus colluded with other representational systems in the society in helping to construct the 'social subjectivity' of contemporary home-makers. *Father Knows Best* and *Leave it to Beaver* thus offered a more naturalistic alternative to slapstick and gag-oriented family sitcom, their plots dealing humorously with the domestic problems of the growing middle-class family within a realistic *mise en scène* that displayed the architecture and accoutrements of the 50s boom.

Haralovich does not study the audience for sitcom as such, nor the historical conditions of TV viewing itself in post-war America (when, as she discusses, the TV set was interpreted not only as a technical apparatus, but as an integral item of domestic furniture). The audience she implies is the gendered audience – particularly the female home-maker – as interpellated not only by the television message, but by the accompanying institutions for social definition and social integration in 50s suburbia.

Identifications
Our second category is concerned with audience 'identifications', focusing on soap opera to analyse the audience's negotiation of popular mass media fictions. Here, Kim Schrøder's cross-cultural reading of the various satisfactions of watching *Dynasty* complements Sonia Livingstone's interpretation, informed by social psychology, of British audiences' perceptual comparisons and contrasts between *Coronation Street* and *Dallas*.

The concern with audience use of popular drama is the focus of Kim Schrøder's essay, 'The Pleasure of *Dynasty*: The Weekly Reconstruction of Self-Confidence', which reports on an empirical analysis of the cross-cultural reception of the American series from interviews conducted with twenty-five American viewers in Los Angeles and sixteen Danish viewers in Copenhagen. Schrøder is less interested in the 'needs' which would preoccupy a more traditional uses and gratifications approach, than with the 'taste culture' which invites a wide range of class and gender subjects to participate in viewing.

He is especially interested in analysis of the ways in which viewers 'commute' between the polarities of 'involvement' and 'distance', or simultaneously occupy those poles. Essential features of involvement

3

are seen as the 'indicative' and 'subjective' realms in which viewers respond to the perceived realism of programme-content and also use programme-content to fantasise possible futures. The space between the two is considered to represent a kind of imaginative threshold; *Dynasty* and its viewing are 'concrete manifestations of liminality and liminal behaviour'.

Other types of involvement described by Schrøder include parasocial forms of interaction such as recognition, self-reflexive forms of escape, and hermeneutically framed activities such as narrative prophecy, narrative justification, and the intuition of unknown identities. By contrast, 'distance' may be based on complaints of predictability and implausibility, yet with 'benefits' parallel to those accruing from 'involvement'. Here, gratification may be obtained through successful self-exposure to *Dynasty* as a weekly test of cultural discrimination, or, more complexly, in that explicit aloofness 'may make it easier for viewers to legitimate their subconscious indulgence in the excessive immorality or the existential anguish of the serial'.

The identifications between the television audience and fictional characters in programmes such as *Coronation Street* and *Dallas* is the focus for Sonia Livingstone's research into 'Viewers' Interpretations of Soap Opera: The Role of Gender, Power and Morality'. Her findings, based upon audience-testing, interestingly bring together the two hitherto remote disciplinary focuses of psychology and textual analysis.

In the first phase of the research major characters in *Dallas* and *Coronation Street* were the subject of a sorting exercise in which separate groups of volunteers classified characters according to perceived similarity of personality. Multi-dimensional scaling then generated a spatial model which positioned the characters according to the most explanatory or basic dimensions underlying judgment.

The second phase involved interpretation of the space by operationalising various theoretical predictions of the oppositions or dimensions. Here, a series of qualities were adduced from Implicit Personality Theory, Gender Stereotyping Theories, and from ideological analysis of soap opera to offer a framework for perceptions of character personality-traits by further groups of respondents for each TV programme.

The responses received related to those theoretical frameworks in revealing ways. Of the major components of Implicit Personality Theory (evaluation, potency, and activity), for example, only potency emerged as a clearly relevant category. Gender findings, for their part, were markedly different in relation to the two programmes, as were perceptions concerning morality. The articulations between the three concerns – power, gender, morality – also varied tellingly between the programmes.

4

Subcultures

Particular audience groupings are the focus of the third and fourth categories of the anthology. Our third pair of essays concentrates on 'Subcultures', with a particular interest in ethnicity. Paula Matabane, in her essay, 'Subcultural Experience and Television Viewing', attempts to specify the social variables affecting the relationship between black urban US television viewers, while Eric Michaels, in his account of 'Hollywood Iconography: A Warlpiri Reading', studies the implications of Aboriginal video usage for unilinear notions of social evolution.

Michaels' Aboriginal audience is an ethnic grouping with 40,000 years of separate development culminating in problematic attempts at integration. A literacy rate of less than 25 per cent reflects, in the case of the Warlpiri, engagement in a non-alphabetic form of graphic design which lies somewhere between the phonological and the pictorial and performs complex 'mapping functions' in a kind of 'metabricolage'. It participates in a complicated patterning of relations between symbol and surface. The images so produced support a conservative cosmology and function as evocative thematics within a strict system of ownership of the image, guarded by conventions of secrecy concerning exhibition and viewing.

It is this cultural language, Michaels suggests, which gives the Warlpiri a relationship to the electronic media which bypasses the literacy phase. Using videorecorders in extended family audiences, the Warlpiri have evolved, in the 1980s, their own practice of 'uses and gratifications'. Broadcast TV is understood in terms of its truth status, through a preoccupation with violence, and in terms of a reversed estimation of the importance of diegetic content and context.

In the several hundred hours of original videotape facilitated by Michaels on the Worth/Adair model, the norms of 'direct cinema' prevail. There is a preference for apparently 'empty' images which, it transpires, convey profound sub-cultural memories for the circle of producers and consumers. Production itself is governed by strong rules concerning the presence of particular 'witnesses' although these may well be debarred from actual presence in the image. As with their graphic design, the Warlpiri then favour video as a field of mnemonic symbols amenable to interpretation when viewed in the proper social and cultural context.

Matabane's sample was composed of 161 black television viewers in Washington DC. Within the relationship between viewing and subculture, she is particularly concerned to understand viewer selectivity, and to plot the relationships between the activity of viewing and experience derived from subcultural routines and location in the social structure.

The equations that generate Matabane's findings involve a number of axes. Television viewing itself is understood through a series of parameters – general television viewing, daily and weekly black character viewing, and the 'perceived reality' of black character programmes. Four definitions govern the last-named – programmes with an all-black setting, with an integrated setting, those typified by urban realism, or by macho fantasy.

Participation in the subculture is measured by race of neighbourhood (coded from census data), tenure of neighbourhood residency, church-going, frequency of church attendance, race of church denomination, community participation and age. Location in the social structure is measured by sex, education, income, subjective social class, home ownership and occupational status.

The bulk of Matabane's essay is devoted to a detailed account of the inter-relations that can be plotted among those variations. She explores the complex categorical links, both within and between those paradigms, and in so doing, generates a sophisticated mapping of socially determined patterns of viewing selectivity within her small chosen sample of the black television audience in North America.

The Child Audience

Children are a key focus for audience studies. Not only are they the group whose interests are regularly invoked when broadcast regulations are constructed, they are also a key target audience for advertisers. In this fourth segment of the volume, Patricia Palmer analyses the forms of social interaction fostered by the activity of television viewing, while Rowell Huesmann and Riva Bachrach offer a detailed study of the differential effects of exposure to TV violence among kibbutz and city children.

Patricia Palmer draws upon symbolic interactionist perceptions for insights into 'The Social Nature of Children's Television Viewing'. Her three-stage study – interviews, observation and questionnaire – conducted with Australian children in Sydney during 1982–83, attempts to combine observation of behaviour in social settings with exploration of the meaning those actions have for the participants.

Palmer's research addresses the pleasurability of TV viewing for children, and the forms of social interaction which support pleasure. She summarises her findings by means of a binary system of comparisons between 'expressive' and 'non-expressive' forms, concluding that, contrary to prevalent assumptions, it is the 'expressive' mode which dominates in children's interaction with the medium.

The construction of a Viewer Activity Index, together with measures for home recreation and for structured leisure, enables Palmer to nuance established correlations between longer hours of viewing and lower socio-economic status. In addition to her contestation of the dominant 'passivity' thesis, she is able to show at this further stage of research that children who watched longer hours of television were also those who claimed to be involved in more activities, irrespective of how those activities are determined by socio-economic status.

Huesmann and Bachrach's essay offers an insightful comparative longitudinal analysis which carefully examines a series of hypotheses about the effect of exposure to television violence in the development of aggression in children. They use the unique bifurcation of Israeli society between city and kibbutz children, giving the research sufficiently different environmental factors to test their hypotheses using a series of statistical tests.

The two groups were interviewed each year over a period of three years about their perception of the realism of certain programmes which the research team had defined as violent in content, and their identification with particular television characters. A variety of measures of behaviour and intellectual development were also collected.

The research shows that television violence viewing is very significantly correlated with aggressiveness among city boys and girls, but not among kibbutz children. Huesmann and Bachrach explain this finding as based on two factors – different viewing opportunities for the two groups, coupled with totally different peer attitudes to aggression. 'In a children's society in which values and norms of behaviour are clear, where accountability to the society is emphasised, where interpersonal aggression is explicitly criticised, and where solo TV and film viewing are infrequent, the children's aggressive behaviour is influenced very little, if at all, by what media violence they do observe.'

Effects

The importance of this statement and conclusion cannot be underestimated. Television culture is but one element of a wider social climate, and to attribute all blame at television's door is illogical.

The 'effects' of television in relation to representations of violence – long a dominant focus for audience research – becomes the central concern of our fifth category. Ian Taylor's social democratic review – and polemic – in relation to the institutional debates on video and violence is accompanied by Barrie Gunter and Jacob Wakshlag's detailed study of the ways in which television viewing can be seen to have effects on perceptions of crime among London residents.

Taylor's review of the left's attitude towards questions of morality and censorship in 'Violence and Video: For a Social Democratic Perspective' initially outlines the campaign and debate in the United

Kingdom about 'video nasties', and confronts the theoretical and methodological problems with beliefs in the effects of television. Central to his essay is a consideration of the liberal-progressive defence of free choice, which, within the existing market arrangements, he sees as nihilistic. The failure to discriminate, he asserts, leads to an unwillingness to speak about the *moral* and the *spiritual* character of some hypothetical, future, non-capitalist society. Taylor calls for 'broad socialist alternatives to imaginary conceptions of the organisation of mortal life and, indeed, of practical matters of health and dying in a socialist society'.

In emphasising the notions of 'family life' and the values of bourgeois respectability as key elements in the 'video nasties' debates, Taylor also returns us to the contested area of the family within broadcasting policies, which is important in many of the debates about television and its audience. Taylor's conclusions also reinforce the importance of a regulated framework for broadcasting policy, and the possibility of arguing for censorship in certain instances.

The complexity of the judgments which have to be exercised by policy-makers and programme personnel are highlighted in Gunter and Wakshlag's essay, 'Television and Perceptions of Crime Among London Residents', which focuses the ways in which audiences perceive the 'real world' and its relationship to their viewing habits.

The research was conducted by means of TV viewing diaries and questionnaires with a London-based viewers panel of the IBA in relation to a single week of programmes in Spring 1985. Diaries were used to record the numbers and kinds of programmes viewed, together with appreciation levels, while the questionnaire investigated personal experience of crime, perceptions of the likelihood of crime and fears of personal victimisation, and competence to deal with an attack.

The study attempted to reflect important enhancements to the cultivation effects model associated with Gerbner's findings in this area. Respondents' TV viewing patterns were thus related to judgments about crime at both personal and societal levels, to perceptions of crime related to particular kinds of environment, local and remote. The specificity of programme content was also assessed.

Gunter and Wakshlag's findings suggest that TV viewing patterns are relatively weak and inconsistent indicators of judgments about crime. Just three viewing variables emerged from the regression analysis as significantly related to perceptions and fear of victimisation: the total amount of TV viewing, US crime-drama viewing, and, perhaps surprisingly, viewing of soap opera.

The elusive character of these interrelations alerts the authors to a more profound complexity. Perhaps indeed the relationship between TV and its viewers is in fact characterised by a strong degree of circularity. Perhaps, for example, greater fear of potential danger in the social environment may encourage people to remain indoors – where they watch more television, and are exposed to programmes which tell them things which in turn reinforce their anxieties?

Critiques

The first phase of the anthology concludes with critiques of received assumptions and methodologies. It includes Rosalind Brunt and Martin Jordin's response to David Morley and the encoding/decoding model, together with a review by Guillermo Orozco-Gomez of debates concerning the cognitive effects of non-educational broadcasting.

The research literature in television studies has produced ideas about the intended and unintended effects of educational broadcasting, but little about the cognitive effects of general output television, argues Guillermo Orozco-Gomez in his essay, 'Research on Cognitive Effects of Non-Educational TV: An Epistemological Discussion'. The concerns and the assumptions which characterise research are major focuses for his analysis of this important gap in educational understanding.

Research, he argues, has been shaped by the economics of the relationship between research groups and the broadcasting industry, producing a pragmatic version of social enquiry associated with the scientistic new American sociology. The drive for objectivity, with its exploitation of increasingly sophisticated techniques for data collection, connects with the production of the individual as the unit of analysis and 'thereby addresses individual differences rather than social or structural characteristics'.

Orozco-Gomez goes on to analyse some of the assumptions which underpin mainstream research. The neutrality of the researcher's questions and purposes, for example, are critiqued through the notion of explanatory relativity and in terms of the structural divergence of explanation and prediction. Assumptions concerning the social neutrality of the television institution are analysed through discussion of the fallacy of technological determinism and of the cultural determination of television form and content.

The proposition that the cognitive effects of television fall predominantly within the area of skills is contested finally in terms of its narrow scope for understanding the complexity of the learning process, and by reference to the fact that the very 'realism' of the television message inevitably affects a broad cognitive domain, including the ideological domain of belief.

The construction of an ethnography of reading is the critical target of Brunt and Jordin's essay, 'Constituting the Television Audience: A Problem of Method'. Their remarks, emerging from a 1984 research project on the role of media in shaping public opinion during the

9

Chesterfield UK parliamentary by-election, challenge the theoretical assumptions which underpin a classic study in the field, David Morley's 1980 *Nationwide* analysis.

The value and distinctions of Morley's project, it is argued, are obscured and undermined by 'the uneasy co-existence of two significantly different approaches to audience study' – on the one hand, a largely qualitative, ethnographic theory, on the other, a largely formalist methodology derived from mainstream quantitative research in the social sciences.

Brunt and Jordin suggest that Morley sets up a false dichotomy between the individually-based interview and group analysis when he sees the latter, the dominant technique for research, as lacking in social contextualisation. This preference also entails, it is argued, problems of representability and generalisability.

Groups are thus understood by Morley not as 'real social entities' with a degree of individuated complexity, but as 'convenient and non-essential methodological devices'. Class thus emerges as a global denominator in place of what Brunt and Jordin would wish to see, in more fully materialist terms, as a range of contextual considerations ranging from the groups' immediate conditions of existence to the broad socio-historical and cultural context.

In his attempts to avoid a mechanical sociologism, it is further argued, Morley produces confused accounts of correspondences and non-correspondences between decodings and the institutional and cultural factors which in turn have material connections with broader social categories such as class. The process of 'decoding' itself, they postulate, requires fuller exploration – beyond the scope of Parkin's tripartite schema of 'dominant', 'negotiated' and 'oppositional' decodings favoured by Morley – as well as the specifics of the television text as a structured system of representation.

Identities and the Economic: Europe
The final essays in the collection relate questions of audience to the political economy of broadcasting. Our penultimate section draws on research into the European situation, including Hans Verstraeten's case study focusing on Belgium and William Melody's study of pan-European television in the era of satellite broadcasting.

The viewer as consumer has been a central figure in recent arguments for the deregulation of broadcasting. The debate, taking place in a number of European countries, has included arguments for and against the continuing validity of public service broadcasting institutions and allied ideologies. It is argued that these broadcasters are out of touch with their audiences, and that a broadcasting market should be created which will respond to viewer demand. Verstraeten's essay, 'Commercial and Public Broadcasting in Belgium:

10

The Tension Between the Economic and the Political Dimension', explores the new broadcasting realities in Europe through a case study of Belgium (to early 1986), showing how the tensions between the economic and political forces are to the fore in the conflicts which have arisen as the public service monopolies have come under scrutiny.

Verstraeten illustrates the specificities of the conflicts in Belgium within the context of different strategies adopted by several organisations of media capital. It is those with international interests who seem to be increasingly dominant and who seek to rationalise constant capital by developing outlets in the consumer market, as part of the information society of the future.

The complicated Belgian situation, exacerbated by the historical divide between Walloon and Flemish interests, and in television terms characterised by a high cable penetration, has resulted in different groupings in each language community marked by a schism between press groups. After exploring these forces Verstraeten examines the political situation which seems to have developed in the wake of increasing multinational power, so that politicians now seek to build images with regard to policies they cannot control. Verstraeten sees the main vulnerability of public broadcasting as its distance from the audience, which will not therefore seek to defend the service against privatisation or a fundamental remodelling.

Verstraeten believes the failure of any left strategies to find a defence of public service broadcasting requires a reconsideration of the left's position towards the future of media provision. His theme is echoed in Professor William Melody's essay, 'Pan-European Television: Commercial and Cultural Implications of European Satellites', delivered as the Opening Address at ITSC 1986. For Melody the key concern, too, is the need to forge a new role for public service broadcasting in an age of transfrontier broadcasting by satellite.

Melody's starting point is the change in programming which will result from a global broadcast market. This, he believes, will replace national services primarily oriented to their domestic market place and will be driven by commercial rather than public service imperatives. His main theme is the implications of pan-European television for the promotion of a European common market and a common European culture.

Melody identifies the difficulties for public broadcasting institutions in adapting to and defining a clear role in the new environment. He is, however, sceptical about the inevitability of the changes: 'The new technologies are tools that have provided particular institutions with the leverage to force major alterations in economic, political and social power relations to the advantage of those institutions applying the technology.'

He stresses the problematic position of the European Commission, which seeks to promote pan-European – but which is, in fact, promoting global – television, which will almost certainly be dominated by US-originated programming. The Commission's need to enhance its power relative to the member states requires the support of non-European transnational corporations for the survival of a pan-European service – a fundamental contradiction. Melody sees the need for citizens' groups to ensure that their interests are not ignored, and believes that an active audience should seek to revive interest in the purpose and scope of public service broadcasting, with a closer accountability of broadcasters and policymakers.

Identities and the Political: Africa and South-East Asia
The problems and strategic responses devised by the different actors in the European market are of little immediate importance in Third World countries, where communications media in general, and broadcasting specifically, have often been seen to have a central and often political role in development policies.

In our final section, Kenya and Malaysia provide case studies in Kyalo Mativo's and Ranggasamy Karthigesu's critical analyses of the role of television in debates over 'development' in the post-colonial Third World.

Kyalo Mativo's essay, 'The Role of Communication in Alternative Development Strategies', stresses the economic importance of the mass media to propagate and justify economic ideas, but his critique questions this role. For Mativo modern communications are unreliable and as inappropriate as many developmental strategies. He believes existing systems of communication should be used to help chart an indigenous path of social development.

Mativo asserts that the central problem of any development policy is the blatant injustice at the heart of the international economic order, within which underdevelopment is endemic. Thus neither communications nor development theories can have universal application. Though communications can act as an economic catalyst in the industrialised countries, the non-indigenous social values they carry in Third World countries endows no certain power, and can cause a multiplicity of conflicts. For instance, in Iran during the Shah's rule, the mass media became uncommunicative, and the unsophisticated traditional systems acquired a high communication content and considerable power. Modern media can thus be inappropriate and irrelevant to development needs, and act to disorient a people's culture.

Traditional communication systems, by which Mativo means folk dances, songs, the market place, etc., can, he believes, accommodate development issues; the only way out of the current quagmire is

12

towards self-reliance, using appropriate technology. His concern is to adapt traditional media and use modern equipment to maintain autonomy, to make popular participation indispensable to the process of change.

Karthigesu's essay, 'Television as a Tool for Nation-Building in the Third World: A Post-Colonial Pattern, Using Malaysia as a Case Study', offers a cautionary tale about broadcasting's role in developing countries. Malaysia is a society with diverse racial groups under the political domination of the Malays, where cultural identities are an important political factor. Television's role in the Third World, he suggests, shadows that of first world television, with the American dominance in programme provision as overwhelming as for Europe or Canada.

For Malaysia, however, there is a different outcome. Since the interests of the Chinese and Tamil minorities have been largely neglected in television since 1969, because of central government control, it is *only* the foreign programmes shown on Radio Television Malaysia (RTM) which unite the audience. Apart from extremely poor provision for the two minority groups in terms of programmes in their own language, and which offer self-representation, the emphasis of local programming is divisive, Karthigesu asserts, because it focuses on the racial divide. Furthermore, Karthigesu seeks to show, government control of broadcasting undermines any possibility of television providing social unification, when it merely acts as an organ to justify government policies.

This tale is a familiar one in many societies, but does pose real problems about the role of television in constructing group identities, and the appropriate models of broadcast organisation. These problems intensify in societies with major ethnic divides, but are as important in the European context, where being black, British and European includes paradoxes and contradictions which cannot be easily addressed by global television.

It also returns us to the key issue of the whole collection – how effective is television in achieving the many ends attributed to it by broadcasters, politicians and pressure groups?

PHILLIP DRUMMOND
RICHARD PATERSON
Directors, ITSC 1986

I MODALITIES

PADDY SCANNELL

Radio Times: The Temporal Arrangements of Broadcasting in the Modern World

Attention to the structuring of time and space must be a central concern of any theory that wishes to take account of the actual conditions that shape and are shaped by the activities and interactions of human beings, as Anthony Giddens has stressed in distinguishing three intersecting planes of temporality in every moment of social reproduction. First, the temporality of immediate experience, the continuous flow of day-to-day life, which I will refer to as *clock time*, bounded by the twenty-four hour day. Second, there is the temporality of the life cycle of living organisms, which I will call *life time*. Third, there is the *longue durée*, the slow, glacial movement of institutional time.[1] I will consider this plane of temporality in broadcasting as calendrical time. All three are inextricably entwined in each other, for the flow of events, in their immediate present circumstances, are always and inescapably implicated both in the 'passing away' of life time and in the motionless *longue durée* that outlasts generations. These different planes of temporality permeate all aspects of broadcast programmes and programming.

In this article I want to pay attention to the unobtrusive ways in which broadcasting sustains the lives and routines, from one day to the next, year in year out, of whole populations, and to reflect on some of the implications of these processes. To do so I will try to account for the ways in which the times of radio and television are organised in relation to the social spaces of listening and viewing. By exploring the relationship between the present moment of programmes in their particularity, and their unceasing iteration in the flow of each day's schedules across the months and years, I will try to render more explicit the connections between the social work of production and reproduction in broadcasting. But this in turn cannot be detached from the special spaces from within which and for which broadcasting produces its programmes and schedules. The places from which broadcasting speaks, and the places in which it is heard and seen, are relevant considerations in the analysis of the communicative contexts that broadcasting establishes as part of the sociable fabric of modern life.

15

The Broadcast Calendar and the National Culture

The calendar is based on natural temporal cycles – the lunar month or solar year – and is a means of regulating, in the long term, the manifold purposes of religious and civil life. It not only organises and coordinates social life, but gives it a renewable content, anticipatory pleasures, a horizon of expectations. It is one means whereby 'the temporality of social life is expressed in the meshing of present with past that tradition promotes, in which the cyclical character of social life is predominant'.[2]

Modern bureaucratic forms of administration coupled with the rigours of factory capitalism demanded new time-keeping habits and work disciplines from whole populations.[3] Older task-oriented work habits with their irregular time patterns were gradually driven out, though the persistence of the feast of St Monday in many trades throughout the nineteenth century shows that older habits died hard. These changes made huge inroads on the many feasts and holidays of the old religious and working year. The new civil calendar was pinched and threadbare in comparison. By 1834 there remained only four annual holidays where eighty years earlier there had been seventeen.[4] This draining away of tradition contributed to that sense of 'meaninglessness' and loss of identity which characterised the experience of modern social life as it was profoundly reshaped in the course of the last century. Modern man 'feels as if he has been dropped from the calendar. The big-city dweller knows this feeling on Sundays'.[5]

Yet the collapse of tradition necessitates its reinvention. And as fast as particular ceremonies and symbols lose their resonance and are relegated to the lumber-room of history, others replace them. In the process of modernisation ritual and tradition shed their intimacy with religion as new secular traditions were rapidly and prolifically invented. Nowhere was this more diligently pursued than in the reconstruction of images and emblems of nationhood.[6]

In Britain, as part of this process, the monarchy was thoroughly revamped and refurbished with a whole new deck of ritual functions and ceremonies.[7] The components of a national culture were beginning to converge in the early twentieth century. There was a national educational system to inculcate, as part of its curriculum, the glories of British history and of English literature.[8] The land itself began to be reclaimed, by the National Trust, as part of the national heritage.[9] Nationalism found musical expression in the Savoy Operas and more profoundly in the music of Elgar.[10] In this period many sports, often of quite recent origin, laid down annual competitions and events. But the full convergence of these developments, their synthesis as elements of a single corporate national life available to all, awaited the establishment of broadcasting in its applied social form and the quite new kind of public – a public commensurate with the whole of society – that it brought into being.[11]

In the course of the 1920s and 1930s BBC engineers arranged thousands upon thousands of outside broadcasts from a wide variety of sources for the growing listening public. They included religious services and sacred music from churches; opera and plays from the theatres and entertainment from variety halls; dance music from cafes and concert music from the concert halls; public speeches by public persons from all sorts of public places; and ceremonies and events that ranged from royal occasions to the song of the nightingale. Added to all these was the coverage of sporting events – football, rugby, cricket, horse racing, tennis, boxing and so on.

In presenting this material the broadcasters did not intervene to restructure it. Most programmes observed real time, the length of the broadcast corresponding to the duration of the event. Radio sought to minimise its own presence as witness, claiming simply to extend the distribution of the event beyond its particular context to the whole listening community. Their appeal, which was very great to an audience unlike today's which takes such things for granted, was that they admitted listeners to public events, to their live presence, in a way no previous technology had been able to do:

> Many of your readers must be office workers. They must know what sort of a life is that of a clerk in a provincial city – a tram-ride to the office, lunch in a tea-shop or saloon bar, a tram-ride home. You daren't spend much on amusements – the pictures and that – because you've got your holidays to think of. We have no Trade Unions and we don't grumble, but it's not an easy life. Please don't think I'm complaining. I'm only writing to say how much wireless means to me and thousands of the same sort. It is a real magic carpet. Before it was a fortnight at Rhyl, and that was all the travelling I did that wasn't on a tram. Now I hear the Boat Race and the Derby, and the opening of the Menai Bridge. There are football matches some Saturdays, and talks by famous men and women who have travelled and can tell us about places.[12]

Such broadcasts unobtrusively stitched together the private and the public spheres in a whole new range of contexts. At the same time the events themselves, previously discrete, now entered into new relations with each other, woven together as idioms of a corporate national life. Nothing so well illustrates the noiseless manner in which the BBC became perhaps *the* central agent of the national culture as the calendrical role of broadcasting; this cyclical reproduction, year in year out, of an orderly and regular progression of festivi-

17

ties, rituals and celebrations – major and minor, civil and sacred – that marked the unfolding of the broadcast year.

The cornerstone of this calendar was the religious year: the weekly observances of the Sabbath through church services and a programme schedule markedly more austere than on other days; the great landmarks of Easter, Pentecost and Christmas; the feastdays of the patron saints of England, Scotland and Wales which occasioned special programmes from the appropriate 'region', though what to do with St Patrick's Day was an annually recurring headache for the programme-makers in Belfast.[13] Bank holidays were celebrated in festive mood while the solemn days of national remembrance were marked by religious services and special feature programmes. Sport of course developed its own calendar very quickly. The winter season had its weekly observances of football, rugby and steeplechasing, climaxing in the Boat Race, the Grand National and the Cup Final. Summer brought in cricket and flat racing, the test matches, Derby Day, Royal Ascot and Wimbledon.

Threaded through the year was a tapestry of civic, cultural, royal and state occasions: the Trooping the Colour, the Ceremony of the Keys, the Lord Mayor's Banquet, the Chairing of the Bard, the Dunmow Flitch, the Shakespeare memorial celebrations at Stratford and much, much more. From the late 1920s onwards programme-makers kept a watchful eye on impending anniversaries as occasions for a potential talk or feature. The two thousandth anniversary of Virgil's death produced a talk on Virgil in English Poetry, while some of the more radical elements conspired to remember republican causes – May Day, the Fall of the Bastille, or the hundredth anniversary of the first great Chartist march.

The broadcast year fell naturally into two: the indoor months of autumn and winter and the outdoor months of spring and summer. One of the first things the radio manufacturers discovered was the seasonal nature of the sale of radio sets which increased sharply as winter came on. Hence the annual trade exhibition, Radiolympia, was held in the autumn as heralding the start of the 'wireless season'.[14] By the late 1920s output was being planned on a quarterly basis, and the autumn quarter was always carefully arranged to woo the fireside listeners with a varied menu of new plays, concerts and variety programmes. The fireside months were generally more well stocked with 'serious' listening matter, but from Whitsun onwards the lighter elements in the programmes were expected to have an increasingly wide appeal. At the same time the broadcasters claimed to have redressed the balance between the seasons of the year, making it possible now to hear good music and plays throughout the summer months when the theatres and concert halls were closed.[15] Thus the programme-planners tried to find broadly appropriate

18

material to suit the climate of the year and the mood and leisure activities of the audience. The highpoint of these activities involved the annual arrangements for Christmas Day.

From the very beginning Christmas was always the most important date in the broadcast year. It was the supreme family festival, an invocation of the spirit of Dickens, a celebration of 'home, hearth and happiness'.[16] It was no coincidence that Reith had worked hard for years to persuade the King to speak, from his home and as head of his family, to the nation and empire of families listening in their homes on this particular day. The royal broadcast (the first was in 1932) quickly became part of the ritual of the British Christmas, and is a classic illustration of that process whereby tradition is reinvented. It set a crowning seal on the role of broadcasting in binding the nation together, giving it a particular form and content: the family audience, the royal family, the nation as family.[17]

Though not all this material recurred with predictable regularity, I have tried to convey a sense of an underlying stable temporal framework for broadcasting, working through the weeks and months of the year. Programme output had a patterned regularity that grew stronger in the National Programme in the course of the 1930s. The effects of this process are incremental. They accumulate in time as they are reproduced through time. In the course of many years, over generations, broadcast output becomes sedimented in memory as traces both of a common past and of the biography of individuals. I would like to illustrate one way in which this process works by an examination of the narrative features of broadcast serials and the ways they connect with the lifetimes of individuals.

Broadcasting, Memory, Lifetime

In a nostalgic essay on the enchanting Russian tales of Leskov, Walter Benjamin laments that the art of storytelling is coming to an end. The communicability of experience is fading away as it is replaced by new forms of communication whose object is information: 'Every morning brings us news from all over the world, and yet we are poor in noteworthy stories. This is because no event any longer comes to us without already being shot through with information. In other words, by now almost nothing that happens benefits storytelling; almost everything benefits information.'[18] I should like briefly to reflect on the implications of this distinction between experience and information in relation to continuous serials and news, and their underlying social functions, as I see them, of remembering and forgetting.

These stories are unlike all others in that they have neither a beginning nor an ending. There is no narrative movement towards resolution and closure. Nor is there any originary point of departure.

19

The first episode of *Coronation Street* or, more recently, *EastEnders*, already presupposes its own narrative world as fully given and in being and motion: the first episode simply 'cuts in' on this world as the last episode 'cuts out' of it. It is rare for these stories to die, but when *Waggoners Walk* (Radio 2) was brought to an untimely end in 1980 by BBC economies, there was no attempt to tie up all the loose ends in the last episode. 'One was left with a sense that the serial had not stopped but was still taking place.'[19] This effect, of a fictional world without end or beginning that exists in parallel with the actual world, is the most powerful and distinctive feature of this kind of narrative, and is the basis of the cumulative pleasures it offers its audience.

A sense of 'unchronicled growth' – of the objective, continuing existence outside the narrative of Ambridge in *The Archers*, or the *Crossroads* Motel – is partly an effect of the way time passes between one episode and another. Between times the characters pursue an 'unrecorded existence' that is resumed in the next episode. In other words, we are aware, Christine Geraghty suggests, that day-to-day life has continued in our absence, an awareness that is powerfully enhanced by the way in which time in the fictional world runs in parallel with the actual world. Serials vary in the scrupulousness with which they observe real time. *The Archers*, Geraghty claims, is the most punctilious in this respect; if a character says it is Thursday it is usually Thursday in the outside world. Both *The Archers* and *Coronation Street* refer regularly to actual events and persons in the real world, and bank holidays, royal weddings and other public events are woven into the programmes as they occur in reality.[20]

It is a strange experience to watch the first ever episode of *Coronation Street*, recorded in grainy black and white back in 1960, and to follow this with an episode in colour from the recent past. There, twenty-five years ago, is Ken Barlow played by William Roache, then just turned twenty-one, an anxious grammar school boy in his first year at college. Now here he is years later and with a chequered career and personal life behind him, an anxious middle-aged, married man. Such an abrupt time lapse immediately makes visible the way in which actor and character (the two are inseparable) have aged over a quarter of a century. And this confirms the truth of the time of the tale as corresponding in all particulars with the movement of lifetime and its passing away. Unlike classic narrative which centres on the person of the hero and on other characters as they impinge more or less directly upon him, the narrative interest in serials is much more evenly distributed among a number of permanent characters whose interrelationships constitute the social world of the story. Again, where classic narrative pursues a single storyline with the odd subplot along the way to a final dénouement, serials maintain several stories running in tandem and overlapping. Thus viewers or listeners are presented with a rich pattern of incident and characterisation in which 'the dramatic is mixed in with the everyday, the tragic with the comic, the romantic with the mundane'.[21] The materials from which these narratives are interwoven are drawn from the recognisably predictable (and occasionally unpredictable) situations and circumstances of interpersonal life: family relations, friendships and enmities, the emotional ups and downs of married life – in short, the stuff of daily life and experience.

To maintain the continuity of the densely textured social world of the narrative calls for unremitting 'backstage' teamwork by the actors and production team. The actors portray the characters whom they play as 'real people', as persons in all particulars. It requires considerable art to bring off this artless effect, and the care with which actors attend to the fine details of the management and maintenance of the self-same identity of the characters they play emerges very clearly from Dorothy Hobson's discussions with the cast of *Crossroads*.[22] The longest running programmes include an archivist in the production team to ensure that no discrepancies appear between past and present, that characters retain consistent biographies, and that birthdays and anniversaries are remembered on the right day from one year to the next. For it is certain that if mistakes creep in they will be spotted by regular listeners or viewers.

When all these things are taken together we can begin to understand how it is that the characters that people these fictional worlds are knowable *in the same way* as people in the actual world are known. This I believe to be the remarkable characteristic of such stories (I cannot think that it is true of any other kind of fictional world in any other medium), and it enables us to account for their well known effect as real and lifelike for their many devotees – for that indeed is what they are. The key is the correspondence between the movement of time in the fictional world and the actual world: for, since these move together and at the same pace, it follows that the lifetime of viewers and listeners unfolds at the same rate as the lives of the characters in the stories. Thus we stand in the same relation to them as we do to our own family, relatives, friends and everyday acquaintances. Moreover access to this fictional world corresponds quite closely to the forms of access we have to people in the real world, and just as in our own lives we acquire, through the years, our own incremental biography as well as an accumulating familiar knowledge of the biographies of those around us, so too, in the same way, we come to know the people in the serials. We can recall past incidents in their lives with the same facility as we can remember events in our own lives and in the lives of those we know and have known for years. And we can drop out of the narratives for a while

21

and later return and pick up the threads in the same way as we resume real-life situations and relationships.

All these points are confirmed, I think, in Dorothy Hobson's sympathetic study of the *Crossroads* audience. Here is Marjory, an elderly widow who lives alone, talking about the marital problems of Glenda in the programme:

> Oh well, that kind of thing, her marriage is not satisfactory, is it! That's the answer to that. Well, I'm going to say something to you. I never thought mine was when I was young. So I can understand how she felt, and it's rather a worry to you. In other words, well of course, in my day they were terribly innocent, weren't they? We didn't really know what was what and what wasn't. And I think we were all a bit frightened. Well, I said tonight, you never felt you could really let yourself go, you were frightened of having a kid and all that sort of thing. Well I can understand that, you see. Now *that* is a thing, when I'm listening to that, I think, 'I can remember I used to be a bit like that'. So what I mean to say, whatever they put in *Crossroads*, it's appertaining to something that could happen in life. It doesn't seem fiction to me.[23]

It can readily be seen that the characters in the programme are spoken of as real people while recognised as fictional. Moreover the connections that Marjory makes between circumstances in the story and in her own life are interactive and reciprocal. It is not the case, as some uses and gratifications studies suggest, that viewers use these stories in some simple way as role-models to guide their own conduct.[24] More interestingly, in moving between both worlds, they use their own life-experience critically to assess the life-world of the story just as much as the reverse.[25]

Such programmes resonate in memory in a double sense: they are both a cultural resource shared by millions, and yet are particular to the lives of individuals. Talk about them is part of the staple currency of the tabloid press and the small change of everyday conversation.[26]

Benjamin's view of stories is essentially passive. They counsel us, we absorb their wisdom. But these audiences have counsel for the people in the stories. Far from dying out in today's societies, the communicability of experience is sustained and augmented by such tales. The repetitiveness of day-to-day life, cross-cut by the irreversibility of lifetime, is continuously reproduced by these stories without beginning or ending.

But the dialectical movement of time always involves more than the continuous presencing of the past in current output. There is also, and necessarily, the simultaneous business of 'letting go' of its

clutches in the forward momentum of change and the turning towards the future. Geraghty puts her finger on much wider issues
when she observes that 'although the accumulated past is important
to a serial, one could also say that the ability to "forget" what has
happened in the serial's past is also crucial. If the serial had to carry
the heavy weight of its own past it would not be able to carry on'.[27]
This work of filtering and discarding the past, while orienting the
present towards emerging and future events and processes, is accomplished daily and routinely by broadcast news.

In a related sense, news production also partakes of a 'time culture'
which emphasises 'the *perishability* of stories'.[28] The whole emphasis
in the newsroom is on 'pace' and 'immediacy'; on news as new and 'up
to the minute'. When a story carries over from one day to the next, it
is assumed that the audience will be familiar enough with the topic
to allow its background to be largely taken for granted. The professional stress is always on *discontinuity*; on upheaval, suddenness,
unpredictability. Thus news lacks any *chronologic*, any continuity
between 'then' and 'now', past and present, which is the mediating
basis of experience.

News then, is the negation of history. It confirms the existence of a
parallel world of public events that is removed from the mundane
day-to-day world of listeners and viewers. The continuous presencing
of this world cannot adequately be accounted for by rationalist explanations of its information function within the mass democratic
politics of nation-states. It is as much, if not more so, a means whereby populations keep 'in touch' with a world external to their own
immediate circumstances because this world can – sometimes in
cruelly unexpected ways – impinge directly upon day-to-day reality.
Over and beyond any particular content, the conventions of daily
news practices are geared to the routines and structures of the
twenty-four hour day, giving it substance as the domain of the
bounded present in our societies.

Broadcasting and the routines of day-to-day life
The critics of mass society and its culture have always chided broadcasting for adjusting to rather than enriching everyday life: 'It takes
the flow of time for granted and is received by audiences who are not
attending particularly to particular moments of communication.
Continuity thus serves to compound the trivialising tendencies of the
universality of the media ... Time, which could have been forcibly
punctuated by broadcasting, has instead been patterned to a rhythm
of acceptability.'[29] In fact in Britain the broadcasters initially
adopted a viewpoint identical to such critics.

In the 1920s and early 1930s, listeners were frequently admonished
to listen attentively and selectively to BBC output. Programme-

23

building (that is, scheduling) was carefully arranged to discourage casual, continuous listening. There were no continuity links between programmes; no announcer's chat, previews of what was to come later, to maintain a smooth flow from one item to the next. Instead the spaces between programmes were left as little oases of silence, save for the tick of a studio clock, to allow people to switch off rather than stay on, or to recompose themselves after a particularly stirring play or concert. Fixed scheduling (placing programmes at the same time on the same day from one week to the next) was, with few exceptions such as the nightly nine o'clock news bulletin, deliberately shunned. It was hard to keep track of, say, a talks series when its day and time varied markedly from week to week – but that was what *Radio Times* was for. Readers were advised to use it as a means of selectively planning their week's listening.

At first, then, the broadcasters suppressed or misunderstood the particular characteristics of their medium and its social implications. They treated radio as an *occasional* resource like theatre, cinema or the concert hall. But radio has no sense of occasion, of a time and place set apart from ordinary life and affairs. On the contrary, its distinctive features are the reverse of occasionality. Because it appears as a domestic utility – always on tap like water, gas or electricity – it must always have an available content. Thus the momentum of broadcasting has always been towards a continuous uninterrupted flow of programme output. Round the clock transmission is the logical completion of this tendency, though its realisation in this country is still held back by economic, cultural and, until the mid-1970s, political constraints.

During the 1930s the BBC was obliged to admit that for most people most of the time, irrespective of class or education, radio was regarded as no more than a domestic utility for relaxation and entertainment – a familiar convenience, a cheerful noise in the background – which in moments of national crisis, mourning or celebration became compulsive listening for the whole population. Modifying the content and structure of the programme service to take account of this discovery entailed greater attentiveness to the circumstances in which listening took place, and to the phased activities of the population through the hours of the day. The pleasures of listening developed on the National Programme were designed to support the new and modest utopia of the suburban nuclear family. The tired businessman or weary office-worker were models frequently invoked to typify the ordinary listener, 'for whom relaxation after a hard day's work was a well earned right. Radio was beginning to play a significant role in the organisation of the rhythm of work and leisure'.[30]

By 1936, under a whole range of pressures, there was less emphasis on cultural enrichment and a greater recognition (reluctantly in some official quarters) that the fireside audience wanted not much more than cheerful, undemanding entertainment. New kinds of programmes were introduced: parlour games and quizzes, light drama serials, and the beginnings of situation comedy with *Bandwagon*. Greater care was taken to organise the daily output on more routinised, regular lines. Popular programmes were increasingly given fixed time slots. *Monday Night at Seven*, as its name implies, was an early and successful attempt to produce a fireside show that recurred at a known time and could pleasurably be anticipated as a predictable enjoyment in the week. By the late 1930s programme-planners were adjusting daily output to chime in with the time routines of day-to-day life through the weekend and the working week.

Today the pattern of output is carefully arranged to match what is known of the daily working, domestic and leisure patterns of the population. Since 1936 when the BBC set up its Listener Research Unit this department has conducted many enquiries into the time use of the British people. Its most recently published study provides a detailed survey of people's time-paths through the day, of when they are in or out of the home, based on diaries kept by nearly 1,400 households from all over the country, selected at random from electoral rolls. The diarists were asked to record the 'main' and 'secondary' activities in half-hour units from 5 a.m. one day to 2 a.m. the next of all household members including children of five and over.[31]

The survey construes time, based on the twenty-four hour clock, as linear through each day and cyclical from one day to the next. Within this cycle three major 'bundles' of time-use are easily discerned. First, necessary or reproductive time for sleep and bodily self-maintenance; second, coercive or work time at office, factory, school or home; third, discretionary or free time for personal activities, leisure and relaxation. The amount of time available in the last category is determined by the prior claims of the first two, which are therefore of great interest to broadcasters who need 'first and foremost to know how many – and what kinds of – people are available to listen or view'.[32]

The diary data was analysed to show what people of different kinds were doing at different times of the day, and to assess how much time people of different kinds spent on different kinds of activity. The results were tabulated in great detail and care is needed in generalising from them. But the overall implication to be drawn, I think, is that the axes of difference both for daily activities and the proportions of time spent on them are determined primarily by age and gender, with occupation and marital status as major determining differences between adults before the age of retirement.[33] Class is not a major point of difference in the overall distribution of these

25

time 'bundles' for the British population, though it is *within* the relatively constant amount of discretionary time and the ways in which it is disposed of. Thus what emerges from the study is the interlocking of lifetime with the time-structures of the day. For what is mapped here synchronically are the different disposals of time-in-the-day by individuals at different stages in their life-cycles. It is *life-position* (a cluster of factors: age, sex, occupation, etc.) that shapes the overall 'time-geography' of people's daily routine.[34]

Broadcasting has moved a long way from its initial demands of careful attention to its programme offerings. Audience research now confirms that listening and viewing has phased patterns of (in)attention through the day.

> During the day, when the radio audience is at its height, the great majority (over 95%) treat listening as only a secondary activity to personal, domestic or work activities. In contrast, viewing audiences reach their maximum during the evening and for over three-quarters of them it is their main activity, only about a tenth doing something else at the same time.[35]

Two moments of the day are of particular interest – the crossover points from home to work in the morning and from work to home in the late afternoon. In these moments of departure and return, how do broadcasters design their programmes to correspond with what they know of the patterns and routines of their intended audiences at such times?

In the early years of television competition both sides observed the so-called 'toddlers truce' – the period between six and seven o'clock each evening when there was no television – partly, it was said, to make it easier for parents to get their young children to bed.

Work on the conditions of reception for cinema has shown it to be a visually overwhelming experience that demands, and gets, a concentrated and psychically charged gaze.[36] By contrast, listening to radio or watching television are distinctively underwhelming experiences. Their usage is interrelated, interactive with other activities and interests, and a means of maintaining sociable contacts between family members, as Bausinger has shown in his account of a typical German household's weekend.[37] The characteristic look that television produces is the glance, and many of its programmes are designed to be understood by people who are only half-watching, popping in and out of the room, or even channel-zapping. Thus broadcasting recognises a distracted listening and viewing subject, pre-eminently figured in the housewife/mother who in Britain tunes in to Radios 1 and 2, and in America to daytime soaps.

In a suggestive account of the relationship between the formal properties of daytime television in the USA and the rhythms of women's work in the home, Tania Modleski argues that the housewife functions by distraction. Daytime television accords closely with the rhythms of women's work in the home and their disrupted, discontinuous work routines. The narrative structures of daytime soaps as well as the flow of other material and commercials tend to make repetition, interruption and distraction pleasurable. Hobson's study of British housewives and their use of radio and television points in the same direction. Domestic work is, in contrast with office or factory routines, essentially structureless. The rolling, continuous daytime popular music programmes on radio provide structures, time boundaries and 'breaks' that help women to organise their routines. The personality of the male DJ provides a point of interaction and release from their situation of 'collective isolation'.

It is beside the point, Modleski argues, to condemn such programmes for distracting the housewife from her 'real' situation. At this level broadcasting and its so-called distractions, along with the particular forms they take, are intimately bound up with women's patterns of work and relaxation.[38]

It is in such instances – and this is but one of many – that production and reproduction converge; that the 'recursive ordering of social life'[39] is routinely sustained by radio and television; that the essential *continuity* of the patterns of day-to-day life are unobtrusively maintained. It is not, of course, that broadcasting creates or determines these patterns, but it is inextricably implicated in them, giving them substance and content, a texture of relevances, presencing in the mundane here-and-now a multiplicity of actual and imaginary worlds, and yet always oriented to, speaking to, the immediate contexts and circumstances of listeners and viewers.

Broadcasting: A World in Common
Throughout this essay I have spoken of *broadcasting*: radio and television together. The privileging of television at the expense of radio in media studies has created a wholly artificial distinction that has distracted attention from the ways in which both are routinely used by populations at different times in the day for different workaday purposes. I have tried to draw attention to this interactive use of both media by people in the phased management of their daily routines, as well as the ways in which the schedules of radio and television, and the form and content of much of their programming, is unobtrusively arranged to fit in with, and to sustain, those routines. Although I have illustrated my arguments with examples drawn mainly from the development of the British system, and of the BBC in particular, I regard the features I have described as characteristic of all national systems of broadcasting in fully developed, modern

industrial societies. It does not matter whether they are organised along public service or commercial lines. Such systems are fundamentally oriented – irrespective of motive or intention – towards the maintenance of the recognisably routine features of day-to-day life for whole populations. At the same time they provide a service attuned to the changing interests, needs and circumstances of people at different phases in their life cycle. Broadcasting, whose medium is time, is profoundly implicated in the temporal arrangements of modern societies.

The fundamental work of national broadcasting systems goes beyond any ideological or representational role. Their primary task is the mediation of modernity, the normalisation of the public sphere and the socialisation of the private sphere. This they accomplish by the continuous production and reproduction of public life and mundane life (nationally and transnationally) not as separate spheres but as routinely implicated in each other, and as recognisable, knowable and familiar. Modern mass democratic politics has its forum in the radically new kind of public sphere that broadcasting constitutes. At the same time radio and television sustain, in individual, interpersonal and institutional contexts, the taken-for-granted accomplishment of all the things we normally do every day of our lives, such as getting out of bed in the morning, washing and dressing ourselves, grabbing a bite to eat and getting off to work or school on time.

I referred earlier to the chronic anxieties produced by the draining away of tradition in the transition to the modern world. The literature of the last century portrayed its mundane world as fraught with tension and danger. For Marx the domain of civil society, and for Dickens the untamed wilderness of the great city, were experienced in the same way: as a hostile environment, 'the sphere of egoism and of the *"bellum omnium contra omnes"* '[40] in which Jonas Chuzzlewit's precept – 'Do other men, for they would do you' – is the best practical advice for survival. Similar themes emerge in Benjamin's study of Baudelaire's work as the poetry of *shock*: a traumatised response to the restless tumult of the metropolitan city. The motif that Benjamin finally selects to encapsulate the essentially modern experience caught by Baudelaire is not that of the *flâneur* (who possesses the streets), but the individual jostled and buffeted by the anonymous crowd – a theme explored by Poe and with which Dickens memorably ended *Little Dorrit*.

If the public world of work and city life seemed strange and threatening, then the private world – the bourgeois and petit-bourgeois home as defence and bulwark against these intrusive pressures – could scarcely prevent those anxieties from penetrating the domestic hearth. The late Raymond Williams acutely caught the social infle-

28

xions of such tensions in the late nineteenth-century European theatre of Ibsen and Chekhov: 'The centre of interest was now for the first time the family home, but men and women stared from its windows, or waited anxiously for messages, to learn about forces "out there", which would determine the conditions of their lives'. The gap between these two worlds, the public and the private, 'the immediate living areas and the directed places of work and government' created both the need and form of new kinds of communication, of 'news' from elsewhere, from otherwise inaccessible sources.[42]

I think of broadcasting's development as a mediating response to such large scale displacements and readjustments in modern industrial societies where custom, tradition and all the givenness of social life has been eroded. Giddens regards routinisation, which he sees as fundamental to any society, no matter how simple or complex, as the basis of 'ontological security' for individual social members. By this he means that individuals have confidence or trust that the natural and social worlds are as they appear to be, including the basic existential parameters of self and social identity.[43] In class-divided nation-states, radio first and later television unobtrusively restored (or perhaps created for the first time) the possibilities of a knowable world, a world-in-common, for whole populations. The social world was rendered sociable, and the manifold anxieties of public life were greatly eased. Broadcasting brought together for a radically new kind of *general* public the elements of a culture-in-common (national and transnational) for all. In so doing, it redeemed, and continues to redeem, the intelligibility of the world and the communicability of experience in the widest social sense.

References

1. Anthony Giddens, *A Contemporary Critique of Historical Materialism* (London: Macmillan, 1981), p. 19.
2. Anthony Giddens, *Central Problems in Social Theory* (London: Macmillan, 1979), p. 201.
3. See E. P. Thompson, 'Time, Work-Discipline and Industrial Capitalism', *Past and Present*, vol. 38, 1963.
4. J. Clarke, C. Critcher, *The Devil Makes Work: Leisure in Capitalist Britain* (London: Macmillan, 1985), p. 56.
5. W. Benjamin, *Illuminations* (London: Fontana, 1973), pp. 186–7.
6. E. Hobsbawm, T. Ranger, *The Invention of Tradition* (Cambridge: Cambridge University Press, 1983), p. 14.
7. See D. Cannadine, 'The Context, Performance and Meaning of Ritual: The British Monarchy and the "Invention of Tradition"', in E. Hobsbawm, T. Ranger, *The Invention of Tradition*.
8. See, respectively, V. Chancellor, *History for Their Masters: Opinion in the English History Text-Book, 1800–1914* (London: Adams and Dart, 1970) and M. Mathieson, *The Preachers of Culture* (London: Unwin Educational, 1975).

9. See M. Bommes, P. Wright, 'Charms of Residence: The Public and the Past', in R. Johnson et al. (eds.), *Making Histories* (London: Hutchinson, 1982).
10. See R. Vaughan Williams, *Natural Music* (London: Oxford University Press, 1963).
11. See P. Scannell, P. Cardiff, 'Serving the Nation: Public Service Broadcasting before the War', in B. Waite et al. (eds.), *Popular Culture, Past and Present* (London: Croom Helm, 1982).
12. Letter in *Radio Times*, 20 January 1928.
13. See R. Cathcart, *The Most Contrary Region: The BBC in Northern Ireland* (Belfast: Blackstaff Press, 1984).
14. See J. Hill, *The Cats Whisker: Fifty Years of Wireless Design* (London: Oresko Books, 1978).
15. See *Radio Times*, 11 May 1934.
16. See *Radio Times*, 20 December 1924.
17. See D. Cardiff, P. Scannell, 'Broadcasting and National Unity', in J. Curran, A. Smith, P. Wingate (eds.), *Impacts and Influences* (London: Methuen, 1987).
18. W. Benjamin, *Illuminations*, p. 89.
19. C. Geraghty, 'The Continuous Serial: A Definition', in R. Dyer, et al., *Coronation Street*, TV Monograph 13 (London: BFI, 1981), p. 11; cf. T. Modleski, *Loving with a Vengeance: Mass Produced Fantasies for Women* (London: Methuen, 1984), p. 90.
20. Geraghty, p. 10.
21. Geraghty, p. 12.
22. See D. Hobson, *'Crossroads'*: The Drama of a Soap Opera (London: Methuen, 1982), pp. 87–105.
23. Quoted in Hobson, pp. 134–5.
24. See M. Cantor, S. Pingree, *The Soap Opera* (Beverly Hills: Sage, 1983), pp. 122–32.
25. See Hobson, pp. 106–37.
26. Cf. I. Penacchioni, 'The reception of popular television in North-East Brazil', *Media, Culture and Society*, vol. 6 no. 4, 1984, and Geraghty, pp. 22–5.
27. Geraghty, p. 18.
28. P. Schlesinger, *Putting 'Reality' Together: BBC News* (London: Constable, 1978), p. 105, my emphasis.
29. P. Abrams, 'Television and Radio', in D. Thompson (ed.), *Discrimination and Popular Culture* (Harmondsworth: Pelican, 1973), p. 113.
30. See S. Frith, 'The Pleasure of the Hearth: The Making of BBC Light Entertainment', in *Formations of Pleasure* (London: Routledge & Kegan Paul, 1983).
31. *The People's Activities and Use of Time* (BBC, 1979). The latest study, *Daily Life in the 1980s* (BBC, 1983) is not generally available. For a brief critical comment on this method, see BBC, *BBC Broadcasting Research Findings*, no. 10 (London: BBC Data Publications, 1984).
32. *BBC, The People's Activities and Use of Time* (London: BBC, 1979), p. 9.
33. For a much fuller discussion of time-budgeting and the BBC surveys, see Science Policy Research Unit, *Changing Patterns of Time Use* (SPRU Occasional Paper no. 13, University of Sussex, 1980).
34. See A. M. Rubin, 'Media Gratifications Through the Life Cycle', in *Media Gratifications Research: Current Perspectives* (Beverly Hills: Sage, 1985), and A. Giddens, *The Constitution of Society* (Oxford: Polity Press, 1984).
35. BBC, *The People's Activities and Use of Time*, p. 522.
36. T. Elsaesser, 'Narrative Cinema and Audience-Oriented Aesthetics' in T. Bennett et al. (eds.), *Popular Television and Film* (London: BFI/Open University, 1981) and J. Ellis, *Visible Fictions: Cinema, Television, Video* (London: Routledge & Kegan Paul, 1982).

37. H. Bausinger, 'Media, Technology and Daily Life', in *Media, Culture and Society*, vol. 6 no. 4, 1984.
38. Modleski, *Loving with a Vengeance*, p. 100.
39. A. Giddens, *Central Problems in Social Theory* (London: Macmillan, 1979), p. 216.
40. K. Marx, *Early Writings* (Harmondsworth: Penguin, 1975), p. 221.
41. Benjamin, p. 195.
42. R. Williams, *Television: Technology and Cultural Form* (London: Fontana, 1974), pp. 26–7.
43. See A. Giddens, *The Constitution of Society* (Oxford: Polity Press, 1984), p. 375.

CLAUS-DIETER RATH

Live/Life: Television as a Generator of Events in Everyday Life

Direct broadcasting, live-TV in the living-room, was the original television mode. The first twenty years of television programming – from 1935 until the late 1950s – were characterised by 'immediate broadcast'. At that time it was impossible to carry out recording, *a posteriori* elaboration and storage of electronic images. If one wanted to copy television programmes, one had to refilm them with a movie camera from the TV screen, which was a very expensive method. To enable repeats, television plays – then broadcast live – had to be re-enacted in front of the camera. It was the development of Ampex video recording (from about 1957) which allowed for a separation in time between the moment of genuine electronic production and its transmission.

Once live broadcasting was no longer a technical necessity it became one choice in certain productions, as well as an integral feature of certain TV genres. Increasingly, live television, which supports the spectator's illusion of being immediately there – of viewing, yet sharing the view via the eye of the camera – was reduced to certain fields. Athletes, announcers, news broadcasters and game-show contestants were 'live'; similarly, parades and scenes of accidents remained subject to this supposedly direct view.

However, the norm for West German television became recorded productions. The stations became distributors of pre-produced material, both from in-house and from purchased programming. Unusual events were witnessed by the audience only when technical difficulties arose or by the intervention of a thunderstorm or the neighbour's kitchen utensils. Sometimes video pirates inserted themselves into the ongoing transmission. Too easily, a camera defect or a badly framed shot could cloud the pleasure of viewing, or an unexpected event or non-event could upset the well planned conception. For management live-television appeared too risky – both technically and from the point of view of content. What could be visibly controlled dominated the process of production.

Direct broadcasting occurs almost in real time, and transmission time is as long as the cameras run. Thus the programme is not

32

established by editing several versions of one take, but instead by the moment-to-moment decisions of the participants. This is why certain things are broadcast which would have been cut out of a recording. Mistakes – or what are considered such – are not eradicable afterwards; the live broadcast is an 'impure' broadcast. However, the opposition live/non-live – that is, direct broadcasting/recording – has nothing to do with the traditional binary oppositions between reality and fiction, or realism and *mise en scène*. Undoubtedly the happening at the scene can be falsified by the broadcast; nonetheless, certain perceptions are possible using television which are not available to the naked eye – such as the view of the earth from the moon or, by means of an endoscope, a look into the inner body. And alongside the whole gamut of camera-work and sound engineering, live-editing also has at its disposal all the electronic means of manipulating image and sound: captions, colour effects, etc.

Since the beginning of the 1980s the big TV stations have also employed portable video cameras, even though use has been restricted to so-called electronic news gathering (ENG). The number of live contributions in regional and local TV news broadcasts increased by exploiting this flexible technique. Live interviews as well as reports in news and current affairs progammes, TV discussions, rock concerts, often broadcast in several countries simultaneously or, exceptionally, to the entire world (for example, *Live Aid*), talk shows and live sport are increasingly present in the programme schedule. The traditional time rituals of broadcasting and viewing are being overridden by live broadcasts. Even such important institutions as the early evening news, which for many people separates the day from the evening, are not spared; it is not uncommon for them to be moved to a different time slot, shortened or extended.

In these cases it is no longer the traditional programme schedule that dominates, but rather the rhythm of events which are thought to be worth broadcasting, even live. Critics complain that important TV plays are now often replaced by live sport. In the third historical phase of TV programming 'live' is being regarded as a new quality for television.

As every medium has its spatial and temporal limits – for example economic efficiency, profitability and the twenty-four hour day – selection has to take place. One of the main criteria for decisions is the topicality of an event. This is the decisive factor which determines the placing and timing of a theme and affects the extent of the audience's attention. Nevertheless, TV producers, when questioned about this, are not able to offer a clear explanation of what 'topicality' is for them.[1]

In essence one can say that topicality is a means of producing attention. The viewer pays attention to something which is marked by special signs and takes note of something made recognisable. Thus it may be assumed that particular topics are of permanent significance in all cultures and for all subjects: love and death, faithfulness and treachery, power and failure, gains and losses.

An object or a process becomes topical when these themes are included at the level of a direct physical stimulus (that is, pain: the 'primary code'), of a linguistic message (that is, a symbolic gesture: the 'secondary code') or of an entire cultural and social order (for example, the collapse or devaluation of certain forms of social exchange: the 'tertiary code'). In this process events from the distant past are brought into a relationship of simultaneity and into the present (for example, 'retro' styles within the fashion industry). The proneness to topicality varies from one society to another. While in static, 'cold' societies topicality is mainly constituted by cyclical events (rites of hunting and of cultivating, of sacrifices, of initiation, of war, etc.), the modern, historical 'hot' societies are found to be permanent generators and consumers of topicality (continual variation of political power constellations, the transport conditions in the cities, daily stock exchange reports, news programmes with special announcements interrupting their rhythm, etc.). But even in modern societies there are differences in this respect: there are relatively 'cold' phases and relatively 'hot' phases (crises, civil war, inflation). Thus we can say that people in modern societies are subjected, in a certain sense, to a permanent initiation procedure through the mode of social topicalities.

If today the journalistic or media-bureaucratic routine is interrupted this happens in the name of a 'topical event'. Events are subjected to the chronological coding of historiographics, which is, as Lévi-Strauss notes, 'never history but the history-for' – for an imaginary other (the future viewer, listener or reader, people from other societies, etc.).[2]

If an event, present or past, is declared to be historical or special, then it must in some way rise above the everyday chaos of details. This particular 'point in time' must be specifically linked to a before and after and to a special content, in order that the discontinuous plurality of facts and the fragments of the 'discrete images in time' suddenly may have a continuity attributed to them.[3] Thus, what has been occurring to a subject is being short-circuited into a fictitious historical stream, whereby the event gains a social and cultural sense. Such a process seems to me to indicate a specific relationship between moment and structure. By its positioning in a continuity the event is integrated into the symbolic order, and thus finds its place in the historical (chronological) code. In this manner the events which are occurring are topicalised and presented as historical. The apparently empty time of everyday life evolves to be totally filled by public and publicised events.

Such kinds of linkage and enrichment of our everyday imagination, reflections and actions are known to us traditionally from participating in festivities or mass events like state events, parades, public trials, executions or mass public outcry, demonstrations or so-called human chains (for example, *Hands Across America* in 1986).

What appears to me to be a decisive element of live television is the relationship between order and its overcoming. Therefore we have to distinguish between the spectator and two levels: the event at the scene with its specific dynamics – catastrophe, quiz, declarations of the state, etc. – and the presentation of the event by the agents of the television institution. Television topicality brings the public into time and this is marked in a specific way. It is not only achieved by the television clock (prior to the news broadcast), which shows the time or (as in quiz games, etc.) the melting away of linear time. This is not merely information as to how late it is (and how time passes) but, rather, whether or not it is too late, or too early, or just the right time to do certain things.

Needless to say, what we get to view is simultaneously being pre-viewed by the editing control monitors and pre-digested, given a rhythm (scanned), and commented upon by camera people and commentators. The particular attraction of live broadcast consists for the viewer in the unforeseen, for instance the Liverpool v. Juventus Cup Final in Brussels and the slaughter in front of the camera. Here, at the outer edge of the symbolic, a moment of horror is constituted. The viewer judges the discourse or the speechlessness of the media agent according to the sights and sounds of that particular scene which are delivered to him. He judges whether or not they, in accordance with the events, succeed in finding the right words, whether they succeed in constructing a *mise en scène* correctly, thus reintegrating the event into the specific symbolic order.

The live broadcast aims at the significant event, which it is trying to capture. This is why it sometimes lasts for an unbearably long time. The climax must be awaited. It is not always possible, however, to ensure an increasing suspense, either via the image or through the intervention of the commentator. Often, nothing is happening, as they say: the soccer match hasn't kicked off yet, the President of the Republic has still not appeared in front of the waiting public to make his declaration, the missile take-off is still postponed, election results are not yet in. But the cameras are transmitting the images, we are there and we stay there, it could be happening in a moment, something, the long awaited unexpected.

The speaker tries to compensate for the non-event by giving additional information in order to maintain interest and suspense. The director nervously switches back and forth from one camera to another, lets the cameras pan and zoom to some supposedly interesting particulars: children, pretty faces, lovers, prominent figures and, when at last nothing else appears to be happening, we see the clouds in the sky. (There is still a lack of television script-writing – and of theory – for the non-event!). Then the apologies begin, first for those viewers who have just turned on their sets and do not understand why nothing is happening. The commentator might let it be known that the producers are considering an interruption of the live transmission – it will be resumed later, now a short film will be shown. After two minutes, however, the latter is interrupted; now it is finally starting, all indications confirm this...

A wide variety of 'media events' are projected into the living-room; partly simulation or deceptions, partly real events whose procedures are oriented towards the television broadcast.[4] Alternatively, small-scale events, inaccessible to the masses and insignificant, become blown up to be media events and are thereby transformed.

There are many different ways for viewers to participate in television from their armchairs. They can express approval or disapproval of a particular topic by simply dialling a certain telephone number the responses to which immediately appear on the screen, or may directly express their own opinion in a television discussion, or by passing on information in a criminal investigation.[5] Sometimes the public may even act directly at the scene. It may be present as a studio audience, or an individual may participate in an entertainment show or ask critical questions in a political feature. Occasionally the audience takes matters into its own hands. This dynamic is well known from accident reports on radio: listeners become curious and rush to the accident scene, thereby blocking rescue services, causing new accidents and making the situation worse.

In the early summer of 1981 in Vermicino, near Rome, a little boy fell into an uncovered well. Italian television reported for two days and nights on three channels without interruption about the rescue work. The rescue attempts failed and the child died, but the whole nation followed the rescue work, which became more and more of a public curiosity. Attracted by the broadcast new actors continuously appeared, claiming to be 'saviours': climbers, pot-holers, acrobats and contortionists all took a chance and were roped down the hole. Each was a part of this macabre live television show.

People don't find themselves by watching television. What happens does not belong to the order of representations by which spectators might feel themselves represented in an adequate way. Viewers recognise themselves as 'socialised' as long as and whenever they are watching television. Since television is not an electronic market place, not an *agora*, not a forum or a parliament, in other words not a public meeting place, it demands neither the participation nor the co-responsibility of the citizen, no active commitment, but a mere being there. The main thing is to remain in the game. What is happening here is often without substance, but is nonetheless, not nothing. What is shown is the world in the specific demonstrative mode of television. One does not gain a lot of information, but rather, the world is spelt out for us. Again and again we are being re-bound in a symbolic order. The normal television aesthetic practises a particular form of show and say which we know from our own childhood: repeatedly things and events are shown and told.

The viewer's attitude is not one of 'I see', but also of 'I also will have seen'. Such a formation of the collectivity around a shared visual perception reminds one of the fan club communities around movie stars, or the readers community around a particular newspaper, or religious communities founded, like the Christian church, on specific books. So, on the occasion of live broadcasts on a global scale via television, for example *Live Aid* in July 1985, the idea of a united humanity can be established and the spectator can feel part of this imaginary totality.

The significance of such collectivity formations is clearly visible in the decision of the Catholic Church, in January 1986, which gave the papal benediction full validity live from television. An interesting comparison can be made with the numerous television church services in the USA and the number of religious stations there. The papal benediction, however, when recorded on video cassette, and preserved and replayable, would seem to lose validity and impact.

References

1. For reflections on the notion of 'topicality', see C.-D. Rath, 'Changes in the Concept of Topicality in the Age of New Electronic Communication Technologies', paper presented to the Commission of the European Communities Conference, 'The Press and the New Technologies: The Challenge of a New Knowledge', Brussels, 1985.
2. C. Lévi-Strauss, *The Savage Mind* (Chicago: University of Chicago Press, 1966), p. 257.
3. G. Simmel, *Soziologie: Untersuchungen über die Formen der Vergesellschaftung* (1908; 5th ed., Berlin: Duncker and Humblot, 1968), pp. 49 and 55.
4. See D. Dayan, E. Katz, 'Rituels publics à usage privé: métamorphose télévisée d'un mariage royal', in *Annales*, vol. 4, 1983.
5. C.-D. Rath, 'The Invisible Network: Television as an Institution in Everyday Life', in P. Drummond, R. Paterson (eds.), *Television in Transition* (London: British Film Institute, 1986).

MARY BETH HARALOVICH

Suburban Family Sitcoms and Consumer Product Design: Addressing the Social Subjectivity of Homemakers in the 1950s

The suburban middle-class family sitcom of the 1950s and 1960s centred on the family ensemble and its homelife – breadwinner father, homemaker mother and growing children, all placed within the domestic space of the suburban home. Structured within definitions of gender and the value of homelife for family cohesion, these sitcoms drew upon particular historical conditions for their realist representation of family relations and domestic space. In the 1950s, a historically-specific social subjectivity of the middle-class homemaker was engaged by suburban housing, the consumer product industry, market research and the lifestyle represented in popular 'growing family' sitcoms such as *Father Knows Best* and *Leave it to Beaver*. With the reluctant and forced exit of women from positions in skilled labour after World War Two and during a period of rapid growth and concentration of business, the middle-class homemaker provided these institutions with a rationale for establishing the value of domestic architecture, materialism and consumer products for an enhanced quality of life and the stability of the family.

The middle-class homemaker was an important basis of this social economy – so much so that it was necessary to define her in contradictions which held her in a limited social place. In her value to the economy, the homemaker was at once central and marginal.[1] She was marginal in that she was positioned within the home, constituting the value of her labour outside of the means of production. Yet she was also central to the economy in that her function as homemaker was the subject of consumer product design and marketing, the basis of an industry. Thus, she was promised psychic and social satisfaction for being contained within the private space of the home; and as a condition of being targeted, measured and analysed for the marketing and design of consumer products, she was promised leisure and freedom from housework.

These social and economic appeals to the American homemaker were addressed to the white middle class whom Stuart and Elizabeth

Ewen have described as 'the landed consumer' for whom 'suburban homes were standardised parodies of independence, of leisure, and, most important of all, of the property that made the first two possible'.[2] The working class is marginalised in, and minorities are absent from, these discourses and the social economy of consumption. An ideal white and middle-class homelife was a primary means of reconstituting and resocialising the American family after World War Two. By defining access to property and home-ownership within the values of the growing family, women and minorities were guaranteed economic and social inequality through racial, ethnic, age and class segregation. While suburban housing provided the kind of gender-specific domestic space and restrictive living environment which was considered therapeutic for the post-war American family, consumer product design and market research directly addressed the class and gender of the homemaker.

The relationship of television programming to the social formation is important to understanding television as a social practice. Graham Murdock and Peter Golding argue that media reproduce social relations under capital through 'this persistent imagery of consumerism conceal[ing] and compensat[ing] for the persistence of radical inequalities in the distribution of wealth, work conditions and life chances.'[3] Stuart Hall has argued that the ideological effects of media fragment class into individuals, masking the determinacy of the economic, and replacing class and economic social relations with imaginary social relations.[4] The suburban family sitcom is dependent upon this displacement of economic determinations onto imaginary social relations which naturalise middle-class life.

Despite its adoption of historical conditions from the 1950s, the suburban family sitcom did not greatly proliferate until the late 1950s and early 1960s. While *Father Knows Best*, in 1954, marks the 'beginning' of popular discussion of the realism of this programme format, it was not until 1957 that *Leave it to Beaver* joined it on the schedule. In the late 1950s and early 1960s, the format multiplied while the women's movement was seeking to release homemakers from this social and economic gender definition.[5] This 'nostalgic' lag between the historical specificity of the social formation and the popularity of the suburban family sitcom on the primetime schedule underscores its ability to mask social contradictions and to naturalise woman's place in the home.

The suburban family sitcom participated in the construction and distribution of social knowledge about the place of women. The following is an analysis of a historical conjuncture in which institutions important to social and economic policies defined women as homemakers: suburban housing, the consumer product industry and market research. *Father Knows Best* and *Leave it to Beaver* mediated this address to the homemaker through their representation of middle-class family life. They appropriated historically specific gender traits and a realist *mise en scène* of the home to create a comfortable, warm and stable family environment. *Father Knows Best*, in fact, was applauded for realigning family gender roles, for making 'polite, carefully middle-class, family-type entertainment, possibly the most non-controversial show on the air waves.'[6]

'Looking through a rose-tinted picture window into your own living-room'

After four years on radio, *Father Knows Best* began the first of its six seasons on network television in 1954. This programme about the family life of Jim and Margaret Anderson and their children Betty (age 15), Bud (13) and Kathy (8) won the 1954 Sylvania Award for outstanding family entertainment. After one season the programme was dropped by its sponsor for low ratings in audience polls. But more than 20,000 letters from viewers protesting the programme's cancellation attracted a new sponsor (the Scott Paper Company) and *Father Knows Best* was promptly reinstated in the primetime schedule. It was a popular programme even after first run production stopped in 1960 when its star, Robert Young, decided to move on to other roles. Reruns of *Father Knows Best* were on primetime for three more years.

Contemporary writing on *Father Knows Best* cited as its appeal the way it rearranged the dynamics of family interaction in television situation comedies. Instead of the slapstick and gag-oriented family sitcom with a 'henpecked simpleton' as family patriarch (presumably programmes like *The Life of Riley*), *Father Knows Best* concentrated instead on drawing humour from parents raising children to adulthood in suburban American. This prompted *Saturday Evening Post* to praise the Andersons for being 'a family that has surprising similarities to real people'.[7] These 'real people' are the white American suburban middle-class family, a social and economic arrangement which was valued as the cornerstone of the American social economy of the 1950s.

The verisimilitude associated with *Father Knows Best* extends not only from the traits and interactions of the middle-class family, but also from the placement of that family within the promises which suburban living and material goods held out for it. Even while the role of Jim Anderson was touted as probably 'the first intelligent father permitted on radio or TV since they invented the thing',[8] the role of Margaret Anderson in relation to the father and the family – as homemaker – was equally important to post-World War Two attainment of quality family life, social stability, and economic growth.

40

Leave it to Beaver was not discussed as much or in the same terms as *Father Knows Best*. Its first run in primetime television was from 1957 to 1963, overlapping the last years of *Father Knows Best*. Ward and June Cleaver raise two sons (Wally, 12 and Theodore – the Beaver, 8) in a single-family suburban home which, in later seasons, adopted an identical floor plan to that of the Andersons. Striving for verisimilitude, the stories were based on the 'real life' experiences of the scriptwriters in raising their own children. 'In recalling the mystifications that every adult experienced when he [sic] was a child, *Leave it to Beaver* evokes a humorous and pleasurably nostalgic glow.'[9]

Like *Father Knows Best*, *Leave it to Beaver* was constructed around an appeal to the entire family as it naturalised middle-class family life. The Andersons and the Cleavers are already assimilated into the comfortable environment and middle-class lifestyle which housing and consumer products sought to guarantee for certain American families. While the Andersons and the Cleavers are rarely (if ever) seen in the process of purchasing consumer products, family interactions are closely tied to the suburban home. The Andersons' Springfield and the Cleavers' Mayfield are ambiguous in their metropolitan identity as suburbs in that the presence of a major city nearby is unclear, yet the communities exhibit the characteristic homogeneity, domestic architecture and separation of gender associated with suburban design.

Margaret Anderson and June Cleaver, in markedly different ways, are two representations of this contradictory function of the homemaker in that they are simultaneously contained and liberated by domestic space. By virtue of their placement as homemakers, they represent the promises of the economic and social processes which were dependent on establishing a limited social subjectivity for homemakers in the 1950s. Yet there are substantial differences in the character traits of the two women, and these revolve around the degree to which each woman is contained within the domestic space of the home. As we shall see, June's self-identity is more suppressed in the role of homemaker than is Margaret's, with the result that June is peripheral in her participation in the family.

Yet these middle-class homemakers have a comfortable life in comparison to television's working-class homemaker. In *Father Knows Best* and *Leave it to Beaver*, the middle-class assimilation of these families is displayed through deep-focus photography which exhibits the taste of middle-class furnishings and the arrangement of tidy rooms, appliances, and gender-specific functional space with private space for men in dens and workrooms and 'family space' for women in kitchens. Margaret Anderson and June Cleaver have a lifestyle and domestic environment which is radically different from that of their working-class sister, Alice Kramden in *The Honeymooners*. The suburban home and consumer products have presumably liberated Margaret and June from the domestic drudgery which marks the daily existence of Alice.

The middle-class suburban environment is comfortable, unlike the cramped and unpleasant space of the Kramden's New York City apartment. Part of the comedy of *The Honeymooners'* working-class and urban family is based on Ralph and Alice Kramden's continual struggle with their outmoded appliances, lower-class taste and economic blocks to achieving the seemingly easy, natural and comfortable assimilation into the middle class through homeownership and consumer goods. Ralph screams out of the apartment window to a neighbour to be quiet; the water pipe in the wall breaks, spraying plaster and water all around; the Kramden's refrigerator and stove are from another era.

One reason for this comedy of *mise en scène* is that urban sitcoms such as *I Love Lucy* and *The Honeymooners* focused on physical comedy and gags generated by their central comic figures (Lucille Ball and Jackie Gleason) and were filmed or shot live on limited sets before a studio audience.[10] *Father Knows Best* and *Leave it to Beaver*, on the other hand, shifted the source of comedy to the ensemble of the nuclear family as it realigned the roles within the family.

Father Knows Best was praised by *Saturday Evening Post* for its 'outright defiance' of 'one of the more persistent clichés of television script-writing about the typical American family ... the mother as the iron-fisted ruler of the nest, the father as a blustering chowderhead and the children as being one sassy crack removed from juvenile delinquency'.[11] Similarly, *Cosmopolitan* cited the programme for overturning television programming's 'message ... that the American father is a weak-willed, predicament-inclined clown [who is] saved from his doltishness by a beautiful and intelligent wife and his beautiful and intelligent children.'[12]

Instead of building family comedy around slapstick, gags and clowning, the Andersons are the modern and model American suburban family, one whom – judging from contemporary articles about *Father Knows Best* – viewers recognised as themselves. *Saturday Evening Post* quoted letters from viewers who praised the programme for being one the entire family could enjoy and even learn from.[13] In *Cosmopolitan*, Eugene Rodney, the producer of *Father Knows Best*, identified the programme's audience demographics as the middle-class and middle-income family.[14] Even in 1959, during its sixth and last first-run season, *Father Knows Best* was still written about as a programme with strong ties to verisimilitude – enabled by its base in a major movie studio with sets that were standing replications of suburban homes – and of therapeutic value for the American family.

'The home is an image ... of the household and of the household's relation to society'

As social historians Gwendolyn Wright and Dolores Hayden have shown, the development and design of suburban housing are fundamental cornerstones of social order. Hayden argues that 'the house is an image ... of the household, and of the household's relation to society'.[15] The single-family detached suburban home was the norm for the family, whose healthy life as a family would be guaranteed by a nonurban environment, neighbourhood stability and separation of family functions by gender. The suburban middle-income family was the primary locus of this homogeneous social formation.

When President Harry Truman said at the 1948 White House Conference on Family Life that 'children and dogs are as necessary to the welfare of this country as is Wall Street and the railroads', he spoke to the role of home-ownership in transforming the post-war American economy. Government policies supported suburban development in a variety of ways. The 41,000 miles of limited access highways authorised by the Federal Aid Highway Act of 1956 contributed to the development of gender-specific space for the suburban family with its commuter husbands and homemaker mothers. Housing starts became, and still continue to be, an important indicator of the well-being of the nation's economy. And equity in home-ownership is considered to be a significant guarantee of economic security in the later years of life.[16]

But while the Housing Act of 1949 defined as its goal, 'a decent home and a suitable living environment for every American family', the Federal Housing Authority (FHA) was empowered with defining 'neighbourhood character' through segregation by class, race, ethnicity and income. And interior space, as well as neighbourhood homogeneity, was designed with an eye toward gender-specific socialising functions for the family. Hayden argues that the two national priorities of the post-war period – removing women from the paid labour force and building more housing – were conflated and tied to 'an architecture of home and neighbourhood that celebrates a mid-nineteenth century ideal of separate spheres for women and men ... characterized by segregation by age, race, and class that could not be so easily advertised.'[17]

In order to establish neighbourhood stability, homogeneity, harmony and attractiveness, the FHA adopted several strategies. Zoning practices prevented multi-family dwellings and commercial uses of property. The FHA also chose not to support housing for minorities and non-nuclear families by adopting a policy called 'red-lining' in which red lines were drawn on maps to identify the boundaries of changing or mixed neighbourhoods. Since the value of housing in these neighbourhoods was designated as low, loans to build and/or buy houses were considered to be bad risks. In turn, banks and savings and loans refused to issue loans and mortgages for 'red-lined' areas. In addition, the FHA published a 'Planning Profitable Neighbourhoods' technical bulletin which gave advice to developers on how to concentrate on homogeneous markets for housing. The effect was to 'green-line' suburban areas, promoting them by endorsing loans and development at the cost of creating urban ghettoes for minorities.[18]

Wright discusses how the FHA went so far as to enter into restrictive or protective covenants to prevent racial mixing and declining property values. Despite the fact that the Supreme Court ruled in favour of the NAACP's case against restrictive covenants, the FHA accepted written and unwritten agreements in housing developments until 1968.[19] The effect of these government policies was to create homogeneous and socially stable communities with racial, ethnic and class barriers to entry. Suburban tracts excluded minorities and were designed to provide the benefits of homeownership to families defined as working husbands, women homemakers and growing children.

Wright describes 'a definite sociological pattern to the household that moved out to the suburbs in the late 1940s and 1950s': average age of thirty-one in 1950; few single, widowed, divorced and elderly; higher fertility rate than in the cities; 9 per cent of suburban women worked compared with 27 per cent in the population as a whole.[20] Hayden lists the five groups which were excluded from single-family housing in the social policies of the late 1940s: single white women, white elderly working and lower-class, minority men of all classes, minority women of all classes, and minority elderly.[21]

The suburban dream house underscored this homogeneous definition of the suburban family. Domestic architecture was designed to display class attributes and reinforce gender-specific functions of domestic space as part of the process of raising children to be adults. Robert Woods Kennedy, an influential housing designer of the period, argued that the task of the housing architect is 'to provide houses that helped his clients to indulge in status-conscious consumption ... to display the housewife "as a sexual being" ... and to display the family's possessions "as proper symbols of socio-economic class" claiming that (this) form of expression (was) essential to modern family life'.[22] In addition to the value of the home for class and sexual identity, suburban housing was also therapeutic for the family. A popular design for the first floor of the home was the 'open floor-plan' which provided a whole living environment for the entire family. With few walls separating living, dining and kitchen areas, space was open for the family togetherness even if they were in separate rooms.

The housing design, built on a part of an acre of private property with a yard for children, allowed the post-war middle-class family to give their children a lifestyle which was not so commonly available during the depression and World War Two. This domestic haven provided the setting for the socialisation of girls into women and boys into men, a place which the breadwinner father paid for with his labour and which the homemaker maintained with hers. Having been placed in the home by suburban development and housing design, the homemaker's life in the home was the concern of the consumer product industry. The basis of its economy, the homemaker was promised leisure from household drudgery and an aesthetically pleasing interior environment.

'Leisure can *transform her life even if good design can't'*
Like suburban housing, the consumer product industry demonstrated an organised desire to replace the heterogeneity of social life with a middle-class consumer culture. And just as housing design and suburban development were based on establishing the natural place for women in the home, the consumer product industry built its economy on defining the social needs and self-identity of women as homemakers. But this industrial definition of the homemaker underwent significant changes during the 1950s as suburban housing proliferated to include the working class.

There were two significant shifts in the focus of discussions among designers about the role of product design in social life. The first was in 1955 when, instead of focusing on the practical problems of design, the Fifth Annual Design Conference at Aspen drew a record attendance to discuss the theoretical and cultural aspects of design. Topics included the role of design in permitting leisure to be enjoyable and the possibility that mass communications could permit consumer testing of product design before the investment of major capital in the product. Design was no longer simply a matter of aesthetically pleasing shapes, but 'part and parcel of the intricate pattern of twentieth-century life'.[23] The second shift in discussions among designers occurred in early 1958 when *Industrial Design* (a major design trade journal) published several lengthy articles on market research which it called 'a new discipline – sometimes helpful, sometimes threatening – that is slated to affect the entire design process'.[24]

Prior to the prominence of market research, designers discussed the contribution of product design to an aesthetically pleasing lifestyle, to the quality of life, and to making daily life easier. While the homemaker was central to the growth and organisation of the consumer product industry, in December 1957 the editors of *Industrial Design* introduced the journal's fourth annual design review with an article which positions the homemaker as a problematic recipient of the benefits of design.

It first summarised the contribution of design to the leisure obtained from consumer products, claiming that 'traditionally American design aims unapologetically at making things easier for people, at freeing them.' It went on to respond to cynics who questioned whether homemakers *should* be freed from housework, arguing for the potentially beneficial emancipation of the homemaker gained by product design. 'More choice in how she spends her time gives the emancipated woman an opportunity to face problems of a larger order than ever before, and this *can* transform her life, even if good design can't.'[25]

These attempts to equate design aesthetics with leisure for the homemaker were occasionally challenged because they marginalised life-styles other than the middle-class. When Dr Wilson G. Scanlon, a psychiatrist, addressed the 1957 meeting of the Southern New England Chapter of the Industrial Designers Institute, he argued that the act of 'excessive purchasing of commodities [was] a form of irrational and immature behaviour', that new purchases and increased leisure have not put anxieties to rest and that 'acceptance of some eccentricity rather than emphasis on class conformity should make for less insecurity [and] a nation that is emotionally mature'.[26]

Esther Foley, home services editor of MacFadden Publications, 'shocked and intrigued' her audience at the 'What Can the Consumer Tell Us?' panel at the 1955 conference of the American Society of Industrial Designers by discussing working-class homemakers.[27] The flagship magazine of MacFadden Publications was *True Story* which had a circulation of 2,000,000 nearly every year from 1926 to 1963. In addition to the confessional stories in the company's *True Romance, True Experience* and other *True* titles, in the 1950s and 1960s some of MacFadden's publications were called 'family behaviour magazines', appealing to the working-class homemakers who were 'not reached by the middle-class service magazines such as *McCalls* and *Ladies Home Journal*'.[28]

Foley introduced a 'slice of life' into the theoretical discussions of design by showing colour slides of the homes of her working-class readers. She showed 'their purchased symbols – the latest shiny "miracle" appliances in badly arranged kitchens, the inevitable chrome dinette set, the sentimental and unrelated living-room furnishings tied together by expensive carpets and cheap cotton throw rugs'.[29]

While Scanlon complained of the psychological damage to the nation from class conformity through consumerism (an issue the women's movement would soon raise), Foley illustrated the disparity between the working class and an aesthetics of product design articulated for the middle class. These criticisms questioned the value of freeing the homemaker from domestic drudgery, the problems of a self-identity centred on consumerism and the narrowly middle-class identification of homemakers by the consumer product industry. Despite this recognition of the social and economic contradictions embedded in consumer products, the homemaker was a fundamental subject of the growing consumer economy.

In the mid-1950s, *Industrial Design* began to publish lengthy analyses of product planning divisions in consumer product corporations. The journal argued that changes in industrial organisation would be crucial to the practice of design. There were three important issues for designers: how large corporations could summon the resources necessary for analysing consumer needs and habits and thus drive smaller companies out of the increasingly competitive market for consumer products; how product designers must become aware of the role of design in business organisations; and how industrial survival in the area of consumer goods would increasingly depend on defining new needs of the consumer.[30]

This need for the consumer product industry to define the homemaker and, through her, its value to homelife is well illustrated by a 1957 discussion among television set designers on whether to design television sets as pieces of furniture or as functional instruments like appliances. The designers talked about three aspects of this problem: how to define the role and function of television in many aspects of daily life, not solely as part of living-room viewing; how to discover the needs of the consumer in television set design; and the necessity of recognising the role of television set design as part of an industry with a mass market. Whether furniture or appliance, television set design should help the homemaker integrate the set into the aesthetics of interior decoration.[31]

The case for television as furniture was based on better 'taste' on the part of consumers and the rapidly expanding furniture industry. Television set purchases exhibited a trend toward 'good taste' and away from the 18th-century mahogany and borax-modern cabinets. In the previous year (1956), the furniture industry had had its best year yet in terms of sales. Given the saturation of television, which reached 87 per cent by 1960, designers agreed that people were spending more time at home and were more interested in its appearance, replacing existing furniture which was wearing out. Television designers needed to consider how television sets would play an important role in this redecoration of the home and how to help the homemaker make these decisions.

The case for television as an instrument rested on its portability. Recent developments in the technology of television allowed the design of smaller and more lightweight sets which could therefore be easily integrated into outdoor activities (on the patio behind the house) and into the kitchen (on the kitchen counter with a cabinet which would match the colour of the appliances). For cues on how to proceed to fill this consumer need, television set designers suggested looking to the appliance industry, which was already effective in integrating products into complete and efficient packages for the kitchen.[32]

The consumer product design industry was also developing an increased awareness of the significance of the homemaker in the economics of marketing and design. Hayden points out that housework is status-producing labour for the family at the same time that it lowers the status of the homemaker by separating her from public life. The 'psychological conflict' engendered from 'guarantee[ing] the family's social status at the expense of her own ... increases when women ... come up against levels of consumption' which lie outside their abilities for upward mobility.[33] Market research based its strength on its ability to turn these tensions around, to place them in the service of the consumer economy.

'Women respond with favourable emotions to the fresh, cream surface of a newly opened shortening can'
By 1958, the 'feminine voice' of the homemaker was even further enmeshed in expert opinion from the field of consumer science and psychology. With high competition in the consumer product industry, it was no longer adequate to determine the conscious needs of the homemaker through interviewing. Instead, market researchers sought to uncover the unconscious processes of consumption. *Industrial Design* described the market researcher as 'a man with a slide rule in one hand and a copy of Sigmund Freud in the other'. The market researcher codified the unconscious motivations in purchasing and quantified them.[34]

The class and gender related tensions inherent in decisions about which consumer product to buy could be identified through market research and alleviated through design. The status of the home and the identity of the homemaker, two important subjects of this research, were based on the development of suburban housing and its concomitant change in shopping patterns. With the impersonal supermarket replacing the small retailer, market researchers argued that sales talk had to be built into product and packaging. This sales talk included two related aspects of market research: defining consumers' awareness of their social identities and class attributes and designing products which would address social identities and class attributes. Survey research, depth or motivational research and experimental research sought to link design with class and gender characteristics and ultimately to determine how product design could appeal to upward mobility and confirm the self-identity of homemakers.

Survey research helped to correlate the 'social image' of products with their users in order to design products which would attract new groups as well as retain current buyers. The Index of Social Position developed by August Hollingshead of Yale University organised data on consumers into an estimation of their social status in the community. A multi-factor system rated residential position (based on neighbourhood), power position (based on occupation) and taste level (based on education). The total score, he argued, would reveal a family's *actual* place in the community, replacing subjective judgments by interviewers. Survey research such as this tried to eliminate interviewer prejudice and render 'objective' evaluations of class status.[35]

This rating system depends on the definition of the family offered by suburban living: breadwinner father, homemaker mother and growing children. Other types of market research focused on the function of women as homemakers, thus placing the economic responsibility for class status with the father and addressing the mother through emotional connotations associated with homemaking. Depth research looked into the psychic motivations of consumers and, for all its shortcomings, led to changes in the designer's address to the consumer. Proof of these researcher deductions and, presumably, the typicality of homemaker emotions, was provided by IBM data processing equipment which could handle large samples and quantify the results.

Experimental research included projective techniques which would elicit unconscious responses to market situations on the theory that consumers would unconsciously impute to others their own feelings and motivations. These included word-association, cartoons in which the word balloons would be filled in, narrative projection in which a story would be finished, role-playing and group discussions. Perception tests hooked homemakers up to machines which measured the speed with which a package could be identified and how much of the 'message' of the design could be retained. Role-playing at shopping and group discussions in the Institute for Motivational Research's 'Motivational Theater' were 'akin to ... "psychodrama"' in that consumers would reveal product, class and gender related emotions which the researchers would elicit and study. These techniques 'stimulate expression' by 'putting oneself in another's position – or in one's own position under certain circumstances, like shopping or homemaking'.[36]

Some designers complained that this application of science to design inhibited the creative process of design by substituting testable and quantifiable elements for aesthetics. But the fact remained that market research techniques were winning out over more traditional designers. In an address to the 1958 Aspen Conference, sociologist C. Wright Mills charged designers with 'bringing art, science and learning into a subordinate relation with the dominant institutions of the capitalist economy and the nationalist state'.[37] Mills' paper was considered to be 'so pertinent to design problems today' that *Industrial Design* ran it in its entirety rather than publishing a synopsis of its major points as it typically did with conference reports.

Mills complained that design helped to blur the distinction between 'human consciousness and material existence' by providing stereotypes of meaning. He argued that consumer products had become 'the Fetish of human life' in the 'virtual dominance of consumer culture'. Mills charged designers with promulgating 'The Big Lie' of advertising and design, that is, 'we only give them what they want'. He accused designers and advertisers of determining consumer wants and tastes which was characteristic 'of the current phase of capitalism in America ... creat[ing] a panic for status, and hence a panic of self-evaluation, and ... connect[ing] its relief with the consumption of specified commodities'. While Mills did not specifically address the role of television, he did cite the importance of distribution in the post-war economy and 'the need for the creation and maintenance of the national market and its monopolistic closure'.[38]

Television suburban homelife in Springfield and Mayfield
One way in which television distributed knowledge about a social economy based on positioning women as homemakers was through the suburban family sitcom. The signifying systems of these sitcoms invested in the social subjectivity of homemakers put forth by suburban development and the consumer product industry. In their representation of middle-class family life, *Father Knows Best* and *Leave it to Beaver* mobilised the discourses of other social institutions to present the home as a haven for the family and the homemaker as contented with and unharried by the terms of her existence. Realistic *mise en scène* and the character traits of family members naturalised middle-class homelife and masked its social and economic barriers to entry.

The heterogeneity of class and gender which market research analysed is not manifested in either *Father Knows Best* or *Leave it to Beaver*. The Andersons and the Cleavers would probably rank quite well in the Index of Social Position. Their neighbourhoods have large and well-maintained homes; both families belong to their respective country clubs. Jim Anderson is a well-respected insurance agent with his own agency (an occupation chosen because it would not tie him to an office). Ward's work is ambiguous but both men carry briefcases and wear suit and tie to work. They have the income which easily provides their families with roomy, comfortable and pleasing surroundings, attractive clothing, and homemakers who need not work outside the home. Both men are college-educated and the programmes often discuss the children's future college education.

Father Knows Best and *Leave it to Beaver* rarely make direct reference to the social and economic means by which the families attained and maintain their middle-class status. Their difference from other classes is not a subject of these sitcoms. By masking the separation of race, class, age and gender by which suburban neighbourhoods were planned, *Father Knows Best* and *Leave it to Beaver* naturalise the privilege of the middle class. Yet there is one episode of *Leave it to Beaver* from the early 1960s which lays bare its assumptions about what constitutes a good neighbourhood. In doing so, the episode suggests how narrowly the heterogeneity of social life can be defined.

Wally and Beaver visit Wally's smart-aleck friend, Eddie Haskell, who has moved out of his family's home into a boarding house in what Beaver recognises as a 'crummy neighbourhood'. Unlike the design of suburban developments, this neighbourhood has large, older, rambling, two-storey (or more) houses set close together (possibly once single family dwellings, now rental units). Porches are large, the door to one house is left ajar, paper debris is blown about by the wind and is left on yards and a front porch. Two men are working on an obviously older model car in the street, hood and trunk open, tyre resting against the car (no garage); two garbage cans are visible on the sidewalks; an older man in sweater and hat walks along carrying a bag of groceries (no car). On a front lawn, a rake leans against a bushel basket with leaves piled up (a sloppy home); a large two-person canvas-covered lawn swing sits on a front lawn (no backyard); one house has a sign in the yard: 'for sale by owner – to be moved' (no real estate agent?).

Wally and Beaver are uneasy in this neighbourhood, one which is obviously in transition and in which activities (working on cars, maintaining lawns, having garbage picked up) are available for public view. But everyone visible is white. This is a rare example of these sitcoms directly defining good and bad neighbourhoods. What is more typical is the assumption that the homes of the Andersons and the Cleavers are representative of their neighbourhoods.

In different ways, the credit sequences which begin these programmes suggest recurring aspects of suburban living. The opening of *Father Knows Best* begins with a long shot of the Anderson's two-storey home, a fence separating the front lawn from the sidewalk, its landscape including trellises with vines and flowers. A cut to the interior entryway shows the family gathering together. In earlier seasons, Jim, wearing a suit and hat in hand, prepares to leave for work. Margaret, wearing a blouse, sweater and skirt, brings Jim his briefcase and kisses him good-bye. The three Anderson children giggle all in a row on the stairway leading up to the second-floor bedrooms. In later seasons, after the long shot of the house, the Anderson family gathers in the entryway to greet Jim as he returns from work. Margaret, wearing a dress too fancy for housework, kisses him at the doorway as their children join them uniting the family in the home.

The opening credits of *Leave it to Beaver*, on the other hand, gradually evolve from an emphasis on the youngest child to placing him within the neighbourhood and then the family. The earliest episodes open with child-like etchings drawn in a wet concrete sidewalk. Middle seasons feature Beaver walking home along a street with single-family homes set back by manicured and landscaped lawns without fences. In later seasons the Cleaver family leaves their two-storey home for a picnic trip: Ward carries the thermal cooler, June (in a dress, even for a picnic) carries the basket and Wally and Beaver join their parents in their late model car.

In *Father Knows Best* the family circulates around the bread-winner father (star Robert Young) who, variously, leaves or returns home. The credit sequence of *Leave it to Beaver* establishes Barbara Billingsley (June Cleaver) first in the cast line up but acknowledges 'Jerry Mathers as the Beaver' as star as well as the source of family situation comedy. *Leave it to Beaver* de-centres the family around the youngest child whose raising to adulthood provides problems which the older child has already surmounted (or never had) while *Father Knows Best* coheres around the family ensemble.

The narrative space of these programmes is dominated by the domestic space of the home. *Father Knows Best* leaves the home environment much less often than *Leave it to Beaver* which is centred on the youngest child who is often seen at school. This placing of the family within the domestic space of the home is, in large measure, the basis of the ability of these programmes to 'seem real'. With their nearly identical interior design, the Andersons' home in Springfield and the Cleavers' home in Mayfield appropriate the gender-specific space which would ensure healthy family life. During the first seasons of *Leave it to Beaver*, the design of the Cleavers' home resembles an older home design more than a suburban dream house. The kitchen is large and homey with glass and wood cabinets. The rooms are separated by walls and closed doors unlike the more open floor plan and modern all-wood kitchen cabinets which the home will adopt in later seasons.

By the 1960s, the Cleavers are living in the 'open floor plan' of the Andersons, a popular housing design of the 1950s. As you enter the homes, to your far left is the den, the private space of the father. To the right of the den is the stairway leading to the 'quiet zone' of the bedrooms. To your right is the living-room, visible through a wide and open entryway the size of two doors. A similar wide and open doorway integrates the living-room with the formal dining-room. A swinging door separates the dining-room from the kitchen. The deep focus cinematography typical of these sitcoms displays the expanse of living space in this 'activity area'.

While the Cleaver children share a bedroom, it comes equipped with a private bathroom and a portable television set. Ward and June's bedroom is small with twin beds. It is rarely seen, not being a site of narrative activity – which typically takes place in the boys' room or on the main floor of the home. These two small bedrooms belie the scale of the house when it is seen in long shot, unlike the Anderson home which makes more use of the potential of the bedrooms for narrative space, especially the 'master' bedroom.

With its four bedrooms, the Anderson home allows each of the children the luxury of his or her own bedroom. Jim and Margaret's 'master' bedroom is larger than those of their children, has twin beds separated by a night stand and lamp, a walk-in closet, a dressing table, arm chairs and a small alcove. The 'master' bedroom was intended to be private space for parents away from children, but the Anderson children have easy access to their parents' bedroom. The Andersons, however, have only one bathroom. Betty has commented that when she gets married she will have three bathrooms because 'there won't always be two of us'.

The Andersons and the Cleavers also share similarities in the decor of their homes, displaying possessions in a comfortably unostentatious way. Immediately to the left of the Andersons' front door is a large, free-standing grandfather clock; to the right and directly across are bookcases built into the wall and filled with hardbacks. In earlier episodes of *Leave it to Beaver* the books (also hardbacks) are on built-in shelves in the living-room. Later, these books are appropriated by Ward's study, lining the many built-in bookshelves behind his desk.

The two families share similar taste in wall decorations and furnishings. Among the landscapes in large and heavy wood frames on the Cleavers' walls are pictures of sailing vessels and reproductions of 'great art' such as Gainsborough's 'Pinky'. While the Andersons do not completely share the Cleavers' penchant for candelabra on the walls and on tables, their walls are tastefully decorated with smaller landscapes. Curiously, neither house engages in the prominent display of family photographs.

The large living-room in each home has a fireplace. There is plenty of room to walk around furniture which is of a decidedly 'non-modernistic' design, over-stuffed and comfortable or hard wood. The formal dining-room in both homes includes a large wooden table and chairs which can seat six comfortably. It is here that the families have their evening meal. A dresser displays dishes, soup tureens, and the like. The kitchen has its own smaller and more utilitarian set of table and chairs where breakfast is eaten. Small appliances such as a toaster, mixer, and coffee-pot sit out on counters. A wall-mounted rack of paper towels is close to the sink. The Andersons' outdoor brick patio has a built-in brick oven, singed from use.

While both homes establish gender-specific areas for women and men, *Father Knows Best* is less repressive in its association of this space with roles within the family. Both Jim Anderson and Ward Cleaver have studies, private areas for the men. Ward is often doing ambiguous paperwork in his den, the rows of hardbacks behind his desk suggesting his association with knowledge and mental work. June's forays into Ward's den tend to be brief, usually in search of his advice on how to handle the boys. Sometimes, as Ward works on papers, June sits in a corner chair sewing a button on Beaver's shirt. Ward's den is often the site of father-to-son talks. Its doorway is wide and open, revealing the cabinet-model television set which Beaver occasionally watches. While Jim has a similar den, it is much less often the site of narrative action and its door is usually closed.

Workrooms and detached garages are also arenas for male activity: lawn-care equipment which the men of the families use; storage for paint, a place to work on the car or smaller engines. The suburban homemaker does not have an equivalent private space. The family space of the kitchen, living-room and dining-room is the woman's space. In typical episodes of *Leave it to Beaver*, June has encounters with her family in the kitchen while Ward's take place in more of the space of the house. As her sons come home through the kitchen door, June is putting up paper towels, tossing a salad, unpacking groceries or making meals. Margaret, having an older daughter, is often able to turn this family/woman space over to her. She is also more often placed within other spaces of the home: the patio, the attic, the living-room.

Both Margaret and June easily adapt to Robert Woods Kennedy's concern that housing design displays the housewife as a sexual being, but this is accomplished not so much through domestic space as through costumes. June's ubiquitous pearls, stockings and heels, and cinch-waisted dresses are amusing in their distinct contradiction with the realities of housework. While Margaret also wears dresses, or skirts, she tends to be costumed in a more casual manner and sometimes wears a smock when engaged in housework such as putting up curtains. Margaret is also occasionally seen in relatively sloppy clothes suitable for dirty work but marked as inappropriate to her status as a sexual being. Yet, even though both programmes were created around 'realistic' storylines of children growing up within the family, the nurturing function of the home and the gender-specific roles of father and mother are handled very differently in *Father Knows Best* and *Leave it to Beaver*.

By 1960, Betty, whom Jim calls 'Princess', has been counselled through adolescent dating rituals but is shown to have 'good sense' and maturity in her relations with boys. Well-kempt and well-dressed like her mother, Betty can easily substitute for Margaret in household tasks such as meal preparation and handling the younger children. In one episode, Jim and Margaret decide that their lives revolve too much around their children ('trapped', 'like servants') and try to spend a weekend away from the family, leaving Betty in charge. While Betty handles the situation smoothly, Jim and Margaret are happier to continue their weekend at Cedar Lodge with all the children along.

Bud, the son, participates in the excitement of discovery and self-definition outside of personal appearance. A normal boy in the process of becoming a man, he gets dirty at sports and tinkering with engines, replaces blown fuses and handles lawn-care. Unlike Betty, Bud has to be convinced that he can handle dating rituals even though Jim counsels him that this awkward stage is normal and one which Jim himself went through. Kathy's pet name is 'Kitten' and, in distinct difference from her sister, she adopts a tomboy persona and interest in sports. By 1959, *Good Housekeeping* is able to purr: 'Kathy seems to have got the idea it might be more fun to appeal to a boy than to be one. At the rate she's going, it won't be long before [Jim and Margaret] are playing grandparents.'[39]

Danny Peary was also pleased with Kathy's development but for a very different reason: in the 1977 *Father Knows Best Reunion* show, Kathy is an unmarried gym teacher. Peary also felt that *Father Knows Best* was different from other suburban family sitcoms in its representation of women: 'The three Anderson females ... were intelligent, proud, and resourceful. Margaret was Jim's equal, loved and respected for her wisdom'.[40] The traits which characterise Margaret in her equality are her patience, good humour, and easy confidence in her family. Unlike Ward Cleaver, Jim is not so immune from wifely banter.

In one episode, Jim overhears Betty and her friend, Armand, rehearsing a play and assumes they are going to elope. Margaret has more faith in their daughter and good-naturedly tries to dissuade Jim from 'acting like a comic strip father'. In the same episode, Jim and Margaret play Scrabble, an activity which the episode suggests they do together often, and which she regularly wins. During the game, Betty returns home from her date with Armand, saying their goodnight on the front porch. Jim is concerned because the porch light has burned out, thus creating a potentially seductive atmosphere for their parting; Margaret, not the least worried about Betty, remarks that she has been asking Jim to replace the bulb for a month.

In contrast to this easy-going family with character traits which allow for many types of familial interaction, *Leave it to Beaver* tells another story about gender relations in the home. June does not share Margaret's status in intelligence, nor her witty and confident relationship with her husband. She typically defers to Ward's greater sense about raising their two sons. Wondering how to approach instances of boyish behaviour, June positions herself firmly at a loss. She frequently asks, mystified, 'Ward, did boys do this when you were their age?' And Ward always reassures June that whatever their sons are doing (brothers fighting, for example) is a normal stage of development of boys, imparting to her his superior social and familial knowledge.

June's sons pass through her space in the kitchen, relaying pieces of information about school and friends as she engages in domestic activities. The episodes usually centre on the private space of the boys' room where Ward often dispenses advice. Like her sons, June acknowledges the need for his guidance. Unlike Margaret, June is structured on the periphery of narrative, secondary to Ward in the socialisation of their sons and in the passive space of the home. Often misogynist, Ward encourages the boys to adopt this attitude toward their mother and to women in general. Unlike *Father Knows Best*, *Leave it to Beaver* works hard to contain June's potential threat to patriarchy. When June asked why Beaver would appear to be unusually shy about meeting a girl, Ward wondered as well: 'He doesn't know enough about life to be afraid of women.'

While *Father Knows Best* and *Leave it to Beaver* position the role of the homemaker in family life quite differently, both women effortlessly maintain the domestic space of the family environment. In their representation of women's work in the home, *Father Knows Best* and *Leave it to Beaver* show the great ease and lack of drudgery with which Margaret and June keep their homes neat, tidy and spotlessly clean. In any episode, these homemakers can be seen in the daily process of doing housework. June prepares meals, waters plants and dusts on a Saturday morning. She brings in groceries, wipes around the kitchen sink and asks Wally to help her put away the vacuum cleaner (which she has not been shown using). Margaret prepares meals, does dishes, irons, and also waters plants. While June is often stationary in the kitchen or sewing on buttons in the living-room, Margaret is usually moving from one room to another, in the process of invisible and on-going domestic activity.

While one could easily argue that this lack of acknowledgment of the labour of homemaking shreds the fabric of verisimilitude of these sitcoms, leisure from housework was promised to homemakers by the consumer product industry. The deep-focus cinematography which reveals the open floor plan of suburban housing design with its tasteful and middle-class furnishings also reveals the appliances in their kitchens. This realism of *mise en scène* integrates consumer products into the domestic space of their homes, suggesting the means by which housework can be facilitated. Margaret and June are unharried homemakers, easily maintaining the comfortable environment necessary for quality family life.

In exchange for being the economic basis of the growing consumer product industry in the 1950s, the homemaker had the potential for receiving the leisure and environment which the acquisition of consumer products assured. Margaret and June mediate these benefits very easily. They are definitely not women of leisure but women for whom housework is neither especially confining nor exclusively time-consuming. While Margaret and June engage in on-going domestic work, they are also unpressured by housework.

While Margaret is more busy about the house than June is, the visible result of their partially visible labour is the constantly immaculate appearance of their homes and variously well-kempt family members. (The older children are more orderly because they are further along in the process of socialisation than the younger ones). Because characters are more fully drawn in *Father Knows Best* than in *Leave it to Beaver*, Margaret engages in more leisure time than June does. She frequently reads in the evenings, plays Scrabble and goes out with Jim and/or the family. *Leave it to Beaver* is more determined to contain June within domestic parameters, thus positioning her character more closely within the framework of domestic servitude. Yet she also leaves the home for occasional entertainment with Ward and has time to read quietly, but to a lesser extent than Margaret.

Margaret and June are contented as homemakers. The 'real time' to do piles of laundry or the daily preparation of balanced meals is a structured absence of the programmes. The free time which appliances provide them is exhibited in their continual good humour and the quality of their interactions with the family. Unrushed and unpressured, Margaret and June are not so free from housework that they become idle and self-indulgent. They are well positioned within the constraints of domestic activity and the promises of the consumer product industry.

Conclusion
We have seen how the homemaker was positioned at the centre of the post-war consumer economy by institutions which were dependent on defining her social subjectivity as a homemaker. In the interests of family stability, suburban development and domestic architecture

were designed with a particular definition of family economy in mind: a working father who could, alone, provide for the social and economic security of his family; a homemaker wife and mother who would maintain the family's environment; children who would grow up in neighbourhoods undisturbed by heterogeneity of class, race, ethnicity and age.

The limited address to the homemaker by the consumer product industry and market research is easily understood when taken within this context of homogeneity in the social organisation of the suburban family. Defined in terms of her homemaking function for the family and for the economy, her life could only be made easier by appliances. To ensure the display of her family's social status, experts assuaged any uncertainties she may have had about interior decor by designing with these problems in mind. Linking her identity as a shopper and homemaker with class attributes in order to broaden the base of the consumer economy, her deepest emotions and insecurities were tapped and transferred to consumer product design.

In their representation of suburban family life, *Father Knows Best* and *Leave it to Beaver* mask these social and economic determinations of the subjectivity of the homemaker and the social and economic inequalities which they assume. The verisimilitude of *mise en scène*, harmonious family life and easy contentment of Margaret Anderson and June Cleaver with their identities as homemakers conceal these processes of subjectivity. The suburban family sitcom illustrated what the working-class family sitcom could not: the easy and unproblematic achievement of quality family life promised by the state and the post-war economy.

Murdock and Golding argue that media studies 'should derive from, and feed into, the continuing debate on the nature and persistence of class stratification'.[41] I have tried to approach this problem by considering the historical definitions of consumers offered by social institutions which, like television, have a stake in addressing certain class and gender identities. Yet this analysis of the place of television in the social economy cannot be accomplished without attention to resistances to institutional imperatives.

Even as the women's movement in the late 1950s and 1960s was exposing these ideologies and the contradictions imbedded in a social economy which positioned women as homemakers, long-running suburban family sitcoms defined women within the security of home-life: *Father Knows Best* (1954–1963), *Leave it to Beaver* (1957–1963), *The Donna Reed Show* (1958–1966), *The Dick Van Dyke Show* (1961–1966), *Hazel* (1961–1966), *Dennis the Menace* (1959–1963), and *The Adventures of Ozzie and Harriet* (1952–1966). This relationship between social institutions and oppositional positions needs to be explored.

References

1. See J. Mitchell, *Women: The Longest Revolution* (London: Virago, 1984), p. 18.
2. S. and E. Ewen, *Channels of Desire: Mass Images and the Shaping of American Consciousness* (New York: McGraw Hill, 1982), p. 235.
3. G. Murdock, P. Golding, 'Capitalism, Communication and Class Relations', in J. Curran et al. (eds.), *Mass Communication and Society* (London: Edward Arnold/Open University Press, 1977), pp. 12 and 36.
4. S. Hall, 'Culture, the Media and the "Ideological Effect"', in Curran et al., *Mass Communication and Society*.
5. I began this study by considering sitcoms which enjoyed three or more seasons on prime time network TV during the 1950s. While the suburban family sitcom accounted for the majority of these 35 series, the period demonstrates a heterogeneity in its approach to sitcom entertainment.
6. K. Rhodes, 'Father of *Two* Families', in *Cosmopolitan*, April 1956, p. 125.
7. B. Eddy, 'Private Life of a Perfect Papa', in *Saturday Evening Post*, 27 April 1957, p. 29.
8. John Crosby, newspaper critic, quoted in Eddy, 'Private Life of a Perfect Papa'.
9. 'TV's Eager Beaver', in *Look*, 27 May 1958, p. 68.
10. T. Brooks, E. Marsh, *The Complete Directory to Prime Time Network TV Shows 1946–Present* (New York: Ballantine, 1981), pp. 340–41, 352–3.
11. Eddy, 'Private Life of a Perfect Papa'.
12. Rhodes, 'Father of *Two* Families,' p. 126.
13. Eddy, 'Private Life of a Perfect Papa'.
14. Rhodes, 'Father of *Two* Families', p. 127.
15. D. Hayden, *Redesigning the American Dream: The Future of Housing, Work and Family Life* (New York: Norton, 1984), p. 40. See also G. Wright, *Building the Dream: A Social History of Housing in America* (Cambridge, Mass.: MIT, 1981).
16. Hayden, pp. 35, 38, 55; Wright, pp. 246, 248.
17. Hayden, pp. 41–2.
18. Wright, pp. 247–8.
19. Wright, p. 248.
20. Wright, p. 256.
21. Hayden, pp. 55–6.
22. Quoted in Hayden, p. 109.
23. 'The Fifth International Design Conference at Aspen found 500 conferees at the crossroads', in *Industrial Design*, vol. 2 no. 4, August 1955, p. 42.
24. Avrom Fleishman, 'M/R, A Survey of Problems, Techniques, Schools of Thought in Market Research Part One', in *Industrial Design*, vol. 5 no. 1, January 1958, p. 26.
25. 'Materialism, Leisure and Design', in *Industrial Design*, vol. 4 no. 12, December 1957.
26. W. G. Scanlon, 'Industrial Design and Emotional Immaturity', in *Industrial Design*, vol. 4. no. 1, January 1957.
27. 'Eleventh Annual ASID Conference: Three Days of Concentrated Design Discussion in Washington, D.C.', in *Industrial Design*, vol. 2 no. 6, December 1955.
28. T. Peterson, *Magazines in the Twentieth Century* (Urbana: University of Illinois Press, 1964), pp. 255, 298, 301–2.
29. 'Eleventh Annual ASID Conference'.
30. R. T. George, 'The Process of Product Planning', in *Industrial Design*, vol. 3 no. 5, October 1956; see also D. Allen et al., 'Report on Product Planning', in

Industrial Design, vol. 4 no. 6, June 1957; 'Lawrence Wilson' and 'Sundberg-Ferar', in *Industrial Design*, vol. 2 no. 5, October 1955.

31. 'IDI Discusses TV, Styling and Creativity', in *Industrial Design*, vol. 4 no. 5, May 1957.

32. On TV technology and TV set design, see, for example, 'TV Sets Gets Smaller and Smaller', in *Industrial Design*, vol. 4 no. 1, January 1957; 'Redesign: Philco Crops the Neck of the Picturetube to be First with Separate-Screen Television', in *Industrial Design*, vol. 5 no. 6, June 1958; 'Design Review', in *Industrial Design*, vol. 6 no. 9, August 1959.

33. Hayden, *Redesigning the American Dream*, p. 50.

34. A. Fleishman, 'M/R: Part 2', in *Industrial Design*, vol. 5 no. 2, February 1958. While *Industrial Design* recognised Paul Lazarsfeld as an important contributor to market research, the journal did not mention his work in the television industry or his development, for CBS, of the Analyzer, an early instrument for audience measurement.

35. Fleishman, p. 35.

36. Fleishman, pp. 41–2.

37. C. W. Mills, 'The Man in the Middle', in *Industrial Design*, vol. 5 no. 11, November 1958.

38. Ibid., pp. 72–4.

39. 'Jane Wyatt's Triple Threat', in *Good Housekeeping*, October 1959.

40. D. Peary, 'Remembering *Father Knows Best*', in J. Fireman (ed.), *TV Book* (New York: Workman Publishing, 1977).

41. Murdock, Golding, 'Capitalism, Communication and Class Relations', p. 12.

II IDENTIFICATIONS

KIM CHRISTIAN SCHRØDER

The Pleasure of *Dynasty*:
The Weekly Reconstruction of Self-Confidence

> Outsiders from the high culture who visit TV melodrama
> occasionally in order to issue their tedious reports about our
> cultural malaise are simply not seeing what the TV audience
> sees.[1]

The Boston Tea Party
In the 105th episode of *Dynasty* almost all of the Carringtons and the
Colbys have gone off to an oil producers' summit in Acapulco, with
the exception of Krystle. Her meetings in Denver with horse-breeder
and tycoon Daniel Reece (Rock Hudson) are being shadowed by a spy
photographer. In Acapulco, Alexis challenges Blake's rights to some
important Chinese oil leases. There is a vicious verbal duel between
British born Alexis (Joan Collins) and a Blake ally, Ashley Mitchell
(Ali McGraw), whose parting shot to Alexis is a subtle warning not to
go too far: 'Watch out, Alexis, remember the Boston Tea Party!'

This episode was used in interviews with Danish viewers of
Dynasty. In this interview with a working-class man (JH) in his mid-
fifties and his somewhat younger wife (HH), I tried, after about
twenty minutes, to get them to talk about the complicated family
relations in *Dynasty*:

Int.: The family relations in *Dynasty are* quite intricate, aren't
they, as you're saying; let's take Reece and Krystle for instance. Reece was in fact married to, no he was going to marry
Krystle's sister, wasn't he?

HH: Yes ... but she died of an illness.

Int.: Yes, but before she died she gave birth to Reece's child, who is
Sammy Jo, right? [Yeah] – So Reece is the father of Sammy Jo,
who is later married to Stephen ...

JH: It's really a bit far-fetched ...

HH: But that photographer, I'd be fucking pleased to know who he
– or she – is, that's for sure.

JH: I believe it's a woman, it was such a slender glove...
HH: I can't figure that one out...
(...)
HH: I'm not sure, but I think this Dominique may be behind that photographer. I really think she is (...) because I think Dominique and Krystle used to know each other, years back, only we haven't been told about it.
JH: And Ashley has something on Alexis, from years back ... Something about a tea party...
Int.: ...something in the past...
HH: Yes, and what the hell's that tea party supposed to mean, we didn't hear anything about that before!
JH: No, and she even mentioned where it took place ... I forget...
Int.: Boston.
JH: Yes

This piece of dialogue can be used in various ways – for example to indulge the facile pleasure of the educated when faced with vulgar ignorance.

The dialogue can also be used as an illustration of the reason for the preference among Danish viewers for Danish programmes. In national programming they understand allusions to the common cultural heritage, to puns, etc., just as probably no American misses the point about The Boston Tea Party.

In the context of this essay I shall use the couple's misunderstandings to shed light on normal processes of understanding, just as the psycholinguist uses research into 'speech defects' to arrive at conclusions about normal speech production.

The couple arrive at *their* understanding of the sequence of TV discourse by relying on the individualised socio-cultural language codes at their disposal. This fixes the propositional meaning of the sequence and uses their reservoir of 'background knowledge' to arrive at the fictional significance.

Clearly, because of insufficient background knowledge they never realise that they have to activate a metaphorical competence in order to generate the rather subtle contemporary significance of the historical reference. Instead they mobilise another aesthetic resource, which they have built up through nearly a hundred weekly exposures to *Dynasty*. Because of their addiction they master to perfection the programme's generic codes, one of which is the gradual initiation of the audience into contrived 'secret' character relations.

As the interview passage shows, JH and HH are constantly scouting the fictional terrain for such concealed connections, and The Boston Tea Party fits neatly into this generic pattern. The decoding accomplished by these two people would normally be categorised as 'erroneous', but for our purposes what matters is that their 'error' provides an insight into decoding processes in general. Every viewer of *Dynasty* (or any other programme) actualises the meaning of the serial in accordance with similar interpretive resources, some of which are culturally shared while others are idiosyncratic. Such interplay between text and recipient, this give-and-take of the signification process, is what lies behind the familiar expression that viewers 'negotiate' the meaning of *Dynasty*; or, emphasising the role of the reader, that viewers '*make* sense of' the TV text.

As a result, the distinction between 'erroneous' and 'correct' readings becomes rather meaningless; only 'actual' readings count, and such actual readings will vary from one person to another. It follows from this that traditional critical analysis of textual content loses any vestige of validity since the claim of such analyses, and their *raison d'être*, has invariably been that they can reproduce the text as received by 'the audience'. There is no such text. My best analysis of the 'Boston Tea Party' sequence – however 'correct' – would have been contradicted by the reading of 90 per cent of the Danish audience.

Finally, the excerpt can be used as an example of one of the pleasures of watching *Dynasty*. When they follow the serial, these two people are engaging in a continuous fictional jigsaw puzzle, constantly on the look-out for new pieces, imagining what they will look like, tentatively fitting them into gaps in the narrative structure or character relations, and experiencing triumphant gratification when they succeed. Among all the aesthetic magnets built into *Dynasty* the game of the fictional puzzle is one of the strongest, rewarding viewers for imaginative skill with a feeling of 'competence' and 'ingenuity'. For the regular viewer, watching *Dynasty* becomes the weekly reconstruction of self-confidence.

The Dynasty Project

The interview on which the introductory section is based was carried out in the context of the *Dynasty* project – an empirical analysis of the cross-cultural reception of *Dynasty*. Two series of interviews were carried out, one with twenty-five American viewers in Los Angeles, the other with sixteen Danish *Dynasty* regulars in the Copenhagen area.[1] The respondents, recruited by two market research companies, were equally distributed with respect to sex and educational status. The interviews took place in respondents' homes, in most cases with the respondent and myself as the only participants.

Each interview proceeded as follows: after the portable VCR had been set up, between five and fifteen minutes were allowed for mutual familiarisation. We would watch episode 101 of *Dynasty*; the accompanying conversation alternated between ordinary small-talk and more pertinent issues, the latter usually on my initiative. The interview following the screening would continue until I felt that all relevant issues had been covered. This phase lasted between five minutes and over an hour.

The interview set-up was designed to overcome, as far as possible, the inherent awkwardness and to neutralise the strong normative controls when watching a low-brow commercial product, particularly in the Danish context. Hence no visible interview guide was used and I frequently stressed my own addiction to the programme and acted as a populist chameleon, reserving provocative questions for the last five minutes of the interview.

The transcribed interviews have been subjected to two types of qualitative analysis: intuitive close-reading (reported in this essay) followed by a more comprehensive and systematic analysis which depends on the development of a social semiotics of reading. As with so many other ongoing studies of cultural phenomena, this project is interdisciplinary in the sense that theories, methods, and results from many academic disciplines are brought to bear on the object of study. This is an absolute necessity when one has the ambition of grasping even a fraction of the kaleidoscopic range of issues raised in the study of audiences and texts.

More specifically, the *Dynasty* Project is interdisciplinary within media research in the sense that it tries to position itself at the point of intersection of two traditionally hostile research paradigms: the social scientific approach, stressing the quantitative analysis of empirical data, and the cultural studies approach, concerned with the critical, often Marxist, qualitative analysis of 'culture industry' ideology.[2] The project, while anchored in the critical tradition, is impatient with its tendency to indulge in speculative analysis of media texts and therefore attempts to apply the sophisticated qualitative tools developed by this tradition to data collected from the real world of the audience. Furthermore, it has no sympathy with cultural-elitist attitudes to commercial television fiction and to popular culture in general.

The fact that commercial television has to aim for a common denominator of tastes and needs does not inevitably result in hopeless programmes that cater only to the lowest instincts and the tritest emotions. It does mean that television has to be understood as both reflector and moulder of mass culture, that television 'functions as a social ritual, overriding individual distinctions in which our culture engages, in order to communicate with its collective self'.[3]

Television entertainment serves as a symbolic representation of the hopes, dreams, and fears that inhabit the collective subconscious of a culture at any given point in time.[4] Viewers' readings of television programmes can be understood as a cultural barometer that registers changes in the cultural climate and the popular mood: current experiences of national identity, attitudes toward social tensions, the faith (or lack of it) in traditional values and established institutions, reactions to new phenomena emerging on the cultural horizon, etc.

Television is a central forum for collective cultural processes, whether as originator or mediator of cultural reinforcement and innovation. In this respect there is a resemblance between television in modern culture and certain types of ritual in tribal society which function as 'a cultural means of generating variability, as well as of ensuring the continuity of proved values and norms'.[5]

Writing about the so-called 'rites of passage', Victor Turner pays special attention to the so-called 'liminal', or threshold, stage of these rites: 'The essence of liminality is to be found in its release from normal constraints, making possible the deconstruction of the "uninteresting" constructions of common sense, the meaningfulness of ordinary life' (...). Liminality is the domain of the "interesting", or of "uncommon sense".'[6]

Therefore, liminal processes may function as a kind of 'metalanguage' through which a population evaluates its own routine behaviours. It is an institutionalisation of the potential for regenerative renewal. Through liminal processes society juxtaposes its 'indicative' mode of existence with 'subjunctive' modes, 'is' with 'may be' or 'should be', and hence provides for itself the potential for cultural change.

For Turner aesthetic media, or performative arts, are the functional equivalents of liminality in complex societies. They compose 'a reflexive metacommentary on society and history as they concern the natural and constructed needs of humankind under given conditions of time and place'.[7] In complex societies it is the performative arts which accomplish the transition from the 'indicative' to the 'subjunctive' mode. Through the consumption of TV fiction, for instance, one enters a symbolic world whose signifying elements offer a communal release from the normal constraints of daily life into the realm of the imaginary.

Turner sees a great opportunity for analysts of modern culture in the exploration of both high art and popular genres, because they

(...) make statements, in forms at least as bizarre as those of tribal liminality, about the quality of life in the societies they monitor under the guise of 'entertainment' – a term which literally means 'holding between', that is 'liminalizing'.[8]

In similar terms *Dynasty* itself and the process of making sense of it can be understood as concrete manifestations of liminality and liminal behaviour. This perspective, in turn, has important consequences for the way the 'hegemonic effect' of television is conceptualised.

While the concept of 'hegemony' itself has introduced a less deterministic view of omnipotent media ideology by pointing out the pluralistic 'leaks' inherent in all successful hegemony, the concept nevertheless implies that the alleged ideological pluralism is ultimately an illusion. Whatever inner tensions and conflicts find expression in media content, 'dominant ideology' remains unperturbed because the ideological leaks, the token admissions of social shortcomings, are always accommodated or co-opted by it.

However, faced with numerous instances in specific TV programmes of genuinely oppositional perspectives, cultural critics are forced to withdraw to a rather diluted 'core' definition of hegemony:

> What is hegemonic in consumer capitalist ideology is precisely the notion that happiness, or *liberty, or fraternity can be affirmed through the existing private commodity forms, under the benign, protective eye of the national security state. *This ideological core is what remains essentially unchanged and unchallenged in television entertainment* at the same time the inner tensions persist and are even magnified. (emphasis added).[9]

Since this was written we have witnessed the arrival of programmes such as *Hill Street Blues*, *Dallas*, and *Dynasty*, in which even this 'core' does not remain 'unchanged and unchallenged'. Are these programmes not hegemonic then? Probably they are, but their version of hegemony is of a precariously defeatist kind: 'Life is miserable, but there is nothing to be done about it.'

Horace Newcomb rightly criticises the 'unchanged core' view of hegemony, arguing that tensions in mass-mediated texts can be regarded as serious 'challenges, in which the terms of the core are redefined or given varied application'.[10] Such real ideological challenges are possible because television texts are mediated by language, whose signifying processes are subject to constant change and multiple interpretations. Meanings cannot be fixed although there is a constant attempt by dominant ideology to achieve just that.

According to Mikhail Bakhtin's theory of *heteroglossia* a vast number of separate 'languages' according to race, class, region, gender, professional registers, group jargon, etc. are embedded into any given piece of social discourse:

> At any given moment of its historical existence, language is heteroglot from top to bottom: it represents the co-existence of socio-ideological contradictions between the present and the past,

between different epochs of the past, between different socio-ideological groups in the present, between tendencies, schools, circles, and so forth, all given a bodily form. These 'languages' of heteroglossia intersect each other in a variety of ways, forming new socially typifying 'languages'.[11]

The visible/audible units of language are empty, or polysemous, *signifiants* invested with a vast range of *signifiés* each originating in a 'language' of its own in the Bakhtinian sense. The Boston Tea Party episode demonstrates how television can only supply the *signifiants*; viewers provide these with *signifiés*. It follows from this that whatever the intentions, no author or text can control the recipients' experience since in the 'heteroglot' situation there will be as many *signifiés*, or discursive systems, as there are recipients.[12]

In the vast majority of cases the provision of meaning will follow processes of (sub)cultural intersubjectivity. But in many cases they do not, and even when they do, no degree of analytical erudition can invest the critical scholar with the ability to designate the different meanings produced by viewers, let alone determine which is 'dominant' and which is 'aberrant'. The path to intersubjective as well as to subjective readings requires empirical studies to actual reception processes. There exists no shortcut from sedentary textual analysis to the textual experience of actual readers. The *Dynasty* Project has so far done little beyond studying the available maps.

The Pleasure of 'Dynasty'
Why do people get addicted to *Dynasty*, so that week after week they will faithfully seek the diverse gratifications its signifiers trigger? Obviously it has something to do with 'needs'. *Dynasty* fascinates because somehow it meets the individualised sociopsychological needs of the viewer trying to make sense of the human condition in modern society.

So far the attempts within uses and gratifications studies to isolate the needs that cause viewers to select specific media and particular programmes have produced few conclusive findings. This is probably for the simple reason that viewers have no awareness of the source of their needs, only of specific impulses to watch one programme rather than another.[13]

Shifting the emphasis from *need* to *taste*, there has been a tendency in recent years to think of audience patterns as not related in simple ways to demographic structures, but to more arbitrary taste preferences. Thus a 'taste culture' is based on 'an aggregate of similar content chosen by the same people'[14], whereby programme choices are detached from social background.

In many ways the taste culture concept is a useful one and its influence is reflected in the fact that the *Dynasty* Project makes no *a priori* assumptions about class, or gender-specific readings. A 'blind analysis' of each interview is made, inferring experiential patterns between viewers only after all interviews have been analysed.

Despite this agnostic approach to the question of socially patterned experiences of TV content, this analysis is sceptical of the view that taste cultures 'are a consequence in part of the genuine classlessness of some uses of leisure time'.[15] People from diverse backgrounds may choose the same media product (as they do with *Dynasty*), but use it differentially. Thus, even if the culturally and socially stratified audience systems may have become neutralised on the level of exposure patterns they may still be operative on the level of decoding.

In other words, we cannot deduce from the socio-economic position of viewers which interpretive strategies they will adopt *vis-à-vis* a specific programme, but 'position in the social structure may be seen to have a structuring and limiting effect on the *repertoire* of discursive or 'decoding' strategies available to different sectors of an audience'.[16]

To sum up, regular *Dynasty* viewers need and like the programme because somehow its signifying structures stimulate the pleasurable attempt to make sense of the human condition in post-industrial capitalist society. For the individual viewer this attempt evokes specific, though not predestined, social patterns, internalised in his or her socialising environment.

Let me now turn to a number of potential dimensions of pleasure for Danish viewers of *Dynasty*. For several of the respondents the weekly viewing of *Dynasty* on Sundays at 5 p.m. was looked upon as a ritual of joy:[17]

> On Sundays when I get out of bed I look forward to *Dynasty* so much I can hardly wait. It's as if, then one can clean the house before noon and things like that, and then be looking forward to the hour when it begins.

The pleasure of *Dynasty* may stem from various sources: the sexual attractions of male and female characters, the display of incredible wealth, the nostalgia for family harmony, and innumerable others. While wishing to reject none of these, this study has found it useful to analyse the experience of watching *Dynasty* in terms of two central concepts, *involvement* and *distance*. I shall try to demonstrate that the experience of the viewer cannot be confined to just one of the two categories, but rather how every viewer moves back and forth, *commutes*, between these two polar opposites. On the one hand there are those viewers who sustain a general involvement in the programme, interspersed with moments of critical distance to some fictional features. On the other hand, there are those whose basically distanced experience is interspersed with moments of fictional involvement.[18]

'Involvement' is not to be thought of as a uniform experience, but as a range of related experiences ('indicative', 'subjunctive'). In the same way 'distance' will be subcategorised into several subtypes ('predictability', 'implausibility', 'aesthetic intrusion'), while the 'hermeneutic' belongs to both categories.

The conceptual framework of the analysis can thus be represented in the following diagram (although opposing terms should not be regarded as pairs):

involvement	distance
subjunctive	predictability
indicative	implausibility
– – – – – – – – –	aesthetic intrusion
	– – – – – – – – –
hermeneutic	hermeneutic

At the extreme end of involvement are those viewers who are totally unable to distinguish between real life and the fictional universe – if such individuals exist among the adult population. Examples might include those who send flowers to soap opera weddings, or those who warn fictional husbands of their wives' infidelity. There were none of these among the respondents. Those taking the greatest distance would be viewers who are so critical of the programme that they may easily become non-viewers, but who go on watching solely in order to achieve continuous affirmation of their cultural superiority. All respondents were less critical than that. In between are those who commute between distance and involvement, and who sometimes feel alienated from and at other times immerse themselves in the programme.

Involvement
Respondent AB is a viewer who is fascinated by *Dynasty*:

And then they have the same human problems like the rest of us, apart from money problems of course. But they are not at all happy, there is no reason to envy them anything except their money and their servants. What I keep telling myself is how nice it would be to have someone to clean the house and keep things tidy (...)

AB sees the *Dynasty* characters from the perspective of *like-us-ness*. Making this sort of comparison between fiction and real life, based on a fairly rational evaluation, AB's comment on the programme in

69

relation to life belongs to the *indicative* mode of discourse. She involves herself in the characters as if they were real people and as if they had the same needs, intentions, feelings, as her. This indicative involvement may occur either in an explicit comparison as in the previous example, or more implicitly when AB slips imperceptibly from fictional to real problems.

Towards the end of her utterance AB adopts another mode of discourse. She imagines what it would be like to possess some of the Carrington attributes. Identifying with fictional characters she enters the *subjunctive* mode; a moment of 'liminality' in which the conditions of everyday life have temporarily been suspended for the benefit of a daydream about what might be. This daydream is heavily invested with the needs originating in AB's burdensome daily life and reflects her desire for more time for herself.

For AB the experience of *like-us-ness* is so compelling that fact and fiction are merging in her consciousness. She is deeply concerned about Claudia's mental problems in the past and the stigma that keeps sticking to her:

> I understand her situation (...), why must one be stigmatized for the rest of one's life because of a mental problem in the past, and that's what happens in this country [Denmark]: 'So you've been to Saint John's [mental hospital], I see!'

In this way the meaning of *Dynasty* may become heavily dependent on the viewer's Danish experience; the characters are subjected to a Danish framework of understanding in which 'Claudia' merges with any female patient in the Copenhagen mental hospital. Earlier in the interview AB had brought in an even more personal element, again prompted by Claudia's situation:

> I really understand her situation because there are so many mentally ill people in Denmark nowadays, and no-one can feel absolutely safe. My young sister is a nurse, and we may all ... tell me who hasn't taken sedative medicine. And that's why – take a look at life insurances (...).

Thus she rambles on to mention alcoholism and other urgent problems. She may in effect be trying to tell me, however obliquely, that she understands Claudia because she herself has been in a similar situation.

The same kind of analogy between American fiction and Danish reality occurs numerous times, on issues like homosexuality, infidelity, divorce, age preferences for erotic/marital partners, etc. For instance AB predicts that Jeff and Nicki are heading towards divorce, continuing:

70

This is really a mess, but do you know what, it's very easy for us to sit here and laugh at them, saying – but it happens very often in real life too, I mean it happens to ordinary people like us, – who doesn't know of a situation where this happened, right, when the woman had a lover, or the husband did, then you get the problem, it's not uncommon at all.

There is an obvious element of *para-social interaction* in AB's experience of *Dynasty*; she is able to establish a vicarious relationship with fictional characters by dissolving the line between fiction and reality. If we use Noble's distinction between two subtypes within parasocial interaction, some aspects of AB's experience fall into the category of *recognition*. She interacts with the fictional characters as if they were real-life people, serving as an extended kin grouping.[19]

For some other viewers this para-social dimension becomes almost tangible: SC describes how the fictional world is invading reality when in some scenes 'Jeff appears to be standing right here in this room'. Not that she minds very much. . . .

Now and again AB does signal a basic awareness of the aesthetic constructedness and distance of the serial, but in most cases the tenuousness of this awareness is evident. While talking about the daughter, Amanda, I ask her:

Int.: Did you watch when the daughter appeared?
AB: Yes.
Int.: Why did she appear – I didn't see the episode when it happened.
AB: Why she appeared, well, because she had become old enough to begin to wonder who her father was and (. . .).

It seems not to occur to her that my ambiguous question could also be interpreted as a request for information about the narrative constructedness of the serial. AB is so enthralled by the *Dynasty* universe that she instinctively perceives the question as pertaining to the motivations intrinsic to characters.

The opposite happens with GO, a respondent whose reading is dominated by distance:

Int.: Do you think Stephen will succeed in – how shall I put it – saving his marriage and making up his mind about his present sexual ambivalence?
GO: No, I don't think so – no, because that's something they can keep using to keep the thing going, isn't it.

71

Even though my question presupposes involvement in Stephen's torn identity, GO ignores this presupposition and her answer refers to the narrative utility of keeping him that way. It is thus evident that respondents' dominant perspective in reading the TV text has crucial consequences for the way they perceive research questions about that text.

Noble's category of *identification* much resembles the type of involvement dealt with here as 'subjunctive', which has more far-reaching ritual implications. 'Identification' occurs when a viewer puts her/himself 'so deeply into a TV character that one can feel the same emotions and experience the same events as the character is supposed to be feeling'.[20]

In the present context, entering the realm of liminality and adopting a subjunctive mode of discourse occurs whenever a respondent projects him/herself into a fictional character, pretending that 'I am character X'; or identifies with a fictional character, wishing that 'I were more like character X'. These processes are both clearly a playful, or 'ludic' acceptance of commuting between fiction and reality, a sort of psychological time-out in which viewers are relatively free to explore the boundaries of their personalities, to look upon 'what they are' in the light of 'what they might be'.

Thus AB, finding Blake condescending when he and Krystle are having an argument about her desire for a career of her own, imagines how Krystle must be feeling and even proceeds to act for her in the dialogue:

> I think she herself feels that they have grown apart. (...) I think that gradually she may be warming up to something and say, 'We have grown apart from each other, and money is not everything.' And she already proved that when she sold her jewellery.

Undoubtedly, AB is here playing with a female role she has sometimes wanted to adopt. At other moments in the interview subconscious slips-of-the-tongue indicate that for years she has 'subordinated' her own desire for education. While her husband has taken a business degree at evening college she 'has not been allowed' to pursue similar goals, but has had to take care of the children and up to three or four menial jobs, 'so that he could relax more'. All of this information she volunteers without any overt bitterness; at one point she even states that 'I don't feel that my needs are being suppressed in any way.' AB may never actually get to adopt the self-assertive role towards her own husband. But fantasising about revolt when watching *Dynasty* does not necessarily mean that she is merely letting off steam so as to endure patriarchal hegemony the better. The imaginary rebellion may equally well be a rehearsal for the real thing. Only time will tell.

Another remark of hers points in the same direction. AB does not like Alexis, but nevertheless she admires some of her qualities:

> Well, nobody really likes Alexis, but on the other hand I think she is a super-egoist, I mean she always puts herself first, not her children, even though she ... this ... but in a way I think ... one likes her and one would like to be a little tough sometimes, so that one was a little more egoistic and not cared so much for everybody else.

As this passage shows, subjunctive involvement should not be confused with sympathy for a character. Overall, Alexis is certainly not a favourite of AB's. But this does not prevent her from 'trying on' certain aspects of the bitchy personality.

Another respondent, SC, explicitly admits that she involves herself in the *Dynasty* fantasy world:

> When I sit down in front of the TV it's as if I ... during that hour one could say that I immerse myself in that world, in those fine dinners and fine drinks and fine clothes. And when it's over, well then I'm just myself again.

By implication, SC's experience of herself watching *Dynasty* is one of being transported to a state of 'not-myself', that is, a ludic putting on of another identity or other identities. In this sense *Dynasty* offers a carnivalesque marketplace of imaginary personality masks, all yours for the taking.

Why do viewers involve themselves in these fictional processes? What are the rewards of commuting back and forth? AB evaluates *Dynasty* in the following manner:

> (...) it is not that thin, but it is somewhat candylike and ... but ... I relax with it, I am not speculating about other things – which I otherwise do ... one always has some kind of problem.

In other words, she *escapes* from the worries of everyday life, a concept which remained undisputed as the whole truth about the pleasure of popular fiction until empirical studies were undertaken, showing its extremely partial nature.[21]

There is no questioning the fact that AB really experiences this relief from worldly worries when she immerses herself in *Dynasty*. What needs to be explained is how this experience can co-exist with the evidence provided by her reading of *Dynasty* as conceptualised in the interview; that she brings every thinkable everyday problem along in her baggage when she watches it:

AB: (...) all those discussions and worries and problems when you turn on the set (...), nothing but misery and there's nothing ...
Int.: Well, but in a way *Dynasty* is misery too, isn't it?
AB: Yes, it is, but after all that's fiction, isn't it?

Fictionalising problems, it would seem, is in itself experienced as making them disappear *as problems*. The thought processes triggered by *Dynasty* are experienced as pleasurable because AB is not accountable to anyone for solutions (not even to herself); the viewing has no purpose beyond itself.

In addition to fictionalisation as such, *Dynasty* is capable of making viewing even more pleasurable because of the way real-life problems are interwoven with blatantly improbable narrative elements, thereby repeatedly offering viewers the possibility of mentally catapulting themselves into a stance of relief with disbelief: this is too unrealistic to be taken seriously!

Perhaps the paradoxical merging of escapism and everyday problems is best expressed in the concept of '*self-reflexive escape*'. This refers to the fact that upon entering any fictional universe we bring along enough of our real-life identities to effect a synthesis of thoughts and feelings between the two types of existence.[22] Furthermore, the occurrence of urgent problems in TV entertainment makes it easier for viewers to realise that their very personal problems are indeed shared within their culture. A programme like *Dynasty* makes us 'relax', as AB puts its, because it bonds us 'as viewers, via the message, to the reality of our culture, thus lifting the burden of an isolating individualism from our shoulders'.[23]

Distance
Right from the first picture of the title sequence GO is highly critical, pointing out reproachfully how wrong it is to associate Krystle with jewellery, how she hates the signature tune so much that if she watches the titles at all (she often doesn't) she turns down the sound.

On the conscious level GO refuses throughout to let herself be involved in fictional events. She tersely points out the contrived, predictable coincidences, as when Stephen and Luke are caught in a (farewell) embrace by Claudia soon after Stephen has promised her never to see Luke again; she vigilantly registers implausible happenings, as when Jeff and Nicki stumble over a precious treasure in a public churchyard; she is heavily ironic about dramatic clichés, commenting 'how expert they are at falling down the stairs'; and she scoffs at passages of artificial dialogue, as when Daniel Reece helps Krystle to her feet after a fall from a horse:

74

Then she fell off her horse, lying there on the grass, then – ugh! then I thought, do people talk to each other like that? (...) It sounds as if they hardly know each other!

The sequence when Nicki manages to get a dead-drunk Jeff to marry her during their Bolivian treasure hunt simply makes her bridle: 'You noticed how he woke up to find he had been married to her – that's utterly unthinkable – that's absurd!' In other words, GO watches *Dynasty* from a position of cultural superiority and deliberately exposes herself to a product that she only allows herself to experience through a filter of condescension, but which nevertheless appears to gratify her immensely. Watching *Dynasty* is a weekly test of her cultural discrimination, and she invariably passes with honours! The pleasure attained through the maintenance of fictional distance thus leads to the weekly reconstruction of self-confidence.

There are also other benefits accruing from the position of aloofness, because this stance may make it easier for viewers to legitimate their subconscious indulgence in the excessive immorality or the existential anguish of the serial.

GO is a woman with a traditional, fairly puritanical sense of morality, who reacts vociferously against the immorality of *Dynasty*. When Alexis and Dex are having an argument upon her return from Blake Carrington's father's deathbed in Sumatra, Alexis justifies her presence there by referring to her marriage to Blake, causing GO to blurt out intrusively:

But that's none of your [Alexis's] business! You've been divorced from him for ages. And *you* [Dex] have been pretty busy, you've been sleeping with Amanda [Alexis's daughter].

Addressing the fictional characters, reproaching them for their lack of propriety and their tainted morals, GO is overtly treating them as 'real' human beings. Evidently the real addressee is her fellow-viewer, and the real objective is not to affect fictional behaviour, but to signal her disapproval of this behaviour. As she puts it later on: 'It's so messy, I really don't understand how American viewers can tolerate such immorality.' What makes it easier for *her* to tolerate the excessive display of divorce, general promiscuity and incest is precisely the stance of critical disbelief, which determines her conscious perception of *Dynasty*. Her ambivalence between spontaneous curiosity and fictional involvement on the one hand (referring to Jeff in third, second and first person address in less than a minute), and critical distance and moral indignation on the other, is illustrated in the following piece of dialogue, again on Jeff and Nicki's marriage:

75

GO: He can't have got very much out of that. That sobered him up! And how did she get that [ring] – she must have brought it from the States. No! You [Jeff] are not just going to swallow that one, are you? Can I [Jeff] see the marriage certificate, please!

Int.: But why do you think he accepts it, because he does!

GO: Well, it might be some sort of, no I suppose it cannot be some sort of code of honour, because he's been to bed with her, and then he thinks he should marry her. (...)

Int.: Is it because he's such a nice guy?

GO: Well, he's not that nice, after all he's had affairs left and right.

Not surprisingly, she ends up rejecting the scene as 'unrealistic' – 'unless that's the way they do these things in America'.

Thorburn defines TV melodrama as 'a sentimental and artificially plotted drama that sacrifices characterisation to extravagant incident, makes sensational appeals to the emotions of its audience, and ends on a happy or at least a morally reassuring note'.[24] He sees these features as the conventions, or established 'rules' of the genre, indeed as 'the *enabling conditions* for an encounter with forbidden or deeply disturbing materials: not an escape into blindness or easy reassurance, but an instrument for seeing'.[25]

In other words, TV melodrama establishes an aesthetic contract with its viewers. It offers them an opportunity to explore individual and social tensions and to face behaviour which is shocking or threatening to prevailing moral codes. Furthermore, it promises that the experience will end on a note of reassurance and moral acceptability, and be stranded with frequent implausibilities so that viewers can suspend involvement and withdraw to a position of superior distance, should they begin to feel uncomfortably affected by the fictional display of agony and immorality.

Thorburn explicitly excludes soap opera from his analysis because of its never-ending serial nature. However, one may still utilise his hypothesis if one adjusts the analysis accordingly. Having excluded itself from the possibility of final reassurance, *Dynasty* has to intensify its supply of alternative unrealistic elements, as recurring safety valves for the viewers, who know that their longing for a happy ending will not be satisfied.

As they are knowledgeable about the aesthetic conventions of *Dynasty*, they expect a constant occurrence of improbable happenings, a blatant constructedness of plots, marvellous coincidences and operatic characterisations. As a result, viewers are usually not in the least upset by these features – they do indeed consider them as conventionalised 'rules of the game'. With Danish viewers this game is perceived as distinctively American:

AB: Fallon will turn up again, but with a different face or ... just like ... well she may not want to be in the serial any more (...) and then suffer a loss of memory, something like that.

Int.:   Now if that sort of thing happened in *Matador* [popular Danish historical series] – would you find that reasonable?

AB: No, that would be more artificial, I guess.

Int.:   Why does one accept it when it happens in *Dynasty* (...)?

AB: (...) well, when it's American one accepts all sorts of things (...) we like to sit and swallow it.

Int.:   (...) that's quite a coincidence, isn't it?

JH: Yeah, but serials are like that.

Even GO, who takes pride in fastening on occurrences of the unrealistic sometimes adopts a more relaxed attitude to these generic conventions:

Int.:   Why do you think we go along with such frequent substitution of actors (...)?

GO: Well, he [Stephen] didn't want to continue, or he made excessive demands so they had to make a substitution with someone else, well, that's just the way it is.

The pleasure of *Dynasty* thus appears to be generated by a fundamental, convention-determined dynamo of alternating involvement in, and distance from, disturbing moral dilemmas of contemporary society.

In some cases one may even suspect that viewers are not *commuting* between the polar opposites, but that they have one foot in each camp, as it were. The experiences of involvement and distance may be *simultaneous* and interdependent, yet still separate. This complex experience seems to be at the bottom of SC's report of her feelings when Jeff was ill: 'What I didn't like was when Jeff was ill. At that time, I was really looking forward to the episode when he wouldn't be ill any more.' Empathetically she suffers with Jeff, but simultaneously and *because* of her empathy she is looking forward to the episode – which as a connoisseur of *Dynasty* genre rules she knows must come – when he is well again. Positioned inside the narrative *and* outside of it she is also in a perfect position to offer predictions and solutions to narrative puzzles:

Even though one may see the appearance of new characters as stupid, well, it's also exciting! So every time I am looking forward to seeing what happens to them. For instance when this woman Dominique suddenly appeared – what sort of a person would she turn out to be ...

In this field of tension between the consciousness of dramatic excess and the desire for imaginative immersion lies the potential pleasure of approaching *Dynasty* as a hermeneutic challenge.

'Dynasty' – Interminable Hermeneutic Puzzle

A couple of years ago the syndicated American soap opera *Rituals* presented a new ratings ploy: it invited its viewers to join a competition based on the serial's narrative development, offering $100,000 to the viewer who could figure out the victim, perpetrator, and motive of a murder to be committed on the show four weeks later.

As the Boston Tea Party interview excerpt quoted at the beginning of this paper shows, *Dynasty* viewers are spontaneously engaging in analogous conjectural pursuits even without the prospect of such a generous prize. Almost all respondents treat *Dynasty* as an ongoing narrative puzzle. Drawing on the memory bank accumulated during months or years of watching the serial, viewers may derive pleasure and self-confidence from unravelling convoluted plots and calculating the origin or predicting the outcome of puzzling narrative elements, such as the parentage of Alexis's long-lost daughter Amanda.

The phenomena dealt with here under the category of narrative puzzle encompass several subtypes, one of which involves *prophesies* about the immediate fictional future:

JH: When she [Dominique] got on stage in order to sing I said, she'll collapse, you just wait and see!

He probably felt quite pleased when she actually did! On another occasion JH suffered a humiliating defeat when he opposed his wife's hypothesis about Dex having malaria:

HH: Last Sunday, with Dex, when he was shivering like that, what did I tell you [husband], he had better get back home, he's got malaria, then Jorgen says, 'How the fuck do you know that?'. Well that's evident, fool, are you stupid or what! Of course he had malaria.

Another puzzle subtype deals with unknown *identities*:

Int.:   Who sent that photographer, that's the question? It's not at all clear whether it is Alexis or Blake.
HH: Oh, talking about *Dynasty*, have you [interviewer] worked out who it is taking those pictures. It's exasperating not to know (...). It might be something Blake had arranged. Or Alexis in

78

order to revenge herself on Blake or something like that, I considered this possibility too – it could also be Dex, she could have got him to do it.

A third subtype covers hypotheses about what the narrative *justification* of aesthetic puzzles will be. The more familiar viewers are with generic conventions, the better will they be able to predict for instance how Fallon will be brought back into the serial (no-one doubts that she *will* come back). Only viewers with adept imaginations will possess the creative skill required to straddle the gap between critical distance and inventive immersion. In order to succeed they have to construct a plausible addition to the aesthetic construct within the boundaries of the fictional conventions; imaginative immersion is necessary in order to find plausible missing pieces to the narrative puzzles (Fallon's death, loss of memory, plastic surgery), without abandoning the consciousness of constructedness ('Fallon' has grown tired of her role).

The notion of television as 'electronic wallpaper' certainly does not adequately describe the average *Dynasty* viewer's approach to the programme. Particularly in the case of keen puzzle-solvers, the interviews document that viewers follow the programme very regularly (some have only missed three or four out of more than a hundred episodes) and pay meticulous attention to all visual and verbal details. For instance JH notices the 'slender hand' of the spy photographer, leading him to hypothesise that it must be a woman, just as he and his wife register Ashley Mitchell's verbal reference to the Boston Tea Party and immediately start to build hypotheses around it. And SC, surmising that Daniel Reece may be behind the spy photographer, vigilantly observes the behaviour of one of Reece's stable hands: 'One of them appears very ... he seems to act kind of funny every time he sees Krystle.'

The conceptualisation of *Dynasty* as a fictional jigsaw puzzle, requiring viewers to engage in a hermeneutic process in order to make sense of the temporary gaps in the text is analogous to (and probably somehow inspired by) Judith Williamson's analysis of the strategies used by advertisements to involve readers in their signifying processes. Advertisers realise that man is a hermeneutic animal and exploit this knowledge in designing ads which invite our participation by requiring us 'to *do* something, to become involved; it is like a children's game or puzzle'.[26]

When we accept the advertisement's invitation to work out a solution and succeed we get the impression that we have accomplished something and feel 'in control' of the text. *We* break up what at a first glance appears an absurdity, *we* work out the meaning of the pun, *we* make the sense! Really this is an illusion, says Williamson, for although puzzles and jokes require us to fill something in, to decipher the meaning,

> these hermeneutic processes are clearly *not* free but restricted to the carefully defined channels *provided* by the ad for its own decipherment. A puzzle has only one solution. A missing piece in a jigsaw has only one shape, defined by its contingent pieces.[27]

However true this may be it does not prevent the individual, child or adult, from experiencing immense gratification from finally fitting in that difficult 'piece' in a crossword puzzle, jigsaw puzzle, advertisement, or soap opera narrative. And it seems doubtful whether anybody would actually suffer from the delusion of hermeneutic freedom in such cases, that is, be ignorant of the fact that the puzzle was constructed with a specific solution in mind. Therefore the effect of such feelings of hermeneutic competence must be judged to be wholly beneficial, increasing the self-confidence of the sense-maker.

There is another aspect, however, which may be more dubious in its ideological effect:

> (...) many complex psychic processes are involved in the work of the ad, and the significance of absences and puzzles in ads is that they give us the opportunity for a 'conscious' activity that masks these unconscious processes. They present their 'manifest' meaning to us latent, thereby concealing the real 'latent' meaning.[28]

In the context of *Dynasty* viewing it is very possible that concentration on the unravelling of plot obscurities takes attention away from less prominent ideological strands. This may explain why *Dynasty*'s overall feeling of existential meaninglessness never surfaces as a compelling experience in the interviews. In a few instances respondents comment on the despair and anguish endemic to some of the serial characters; but the general meaninglessness of the human condition in the modern world is never raised to the level of manifest articulation. Evidently, this absence of existential comments may stem from the fact that no-one 'reads' such issues from the serial text. Alternatively the reason may be that viewers immerse themselves in hermeneutic puzzles so as to divert their conscious attention away from more disturbing insights into the human predicament. This of course does not preclude the possibility that existential readings are seething in the subconscious of *Dynasty* viewers. It just means that, through fictional puzzles, *Dynasty* offers its loyal audience a built-in safety valve against the compelling possibility that life is meaningless, assisting their instinct, against whatever odds, to *make* sense.

References

1. D. Thorburn, 'Television Melodrama', in H. Newcomb (ed.), *Television: The Critical View* (New York: Oxford University Press, 1979).
2. See T. Gitlin, 'Media Sociology: The Dominant Paradigm', in *Theory and Society*, vol. 6, 1978, and K. Schrøder, 'Convergence of Antagonistic Traditions? The Case of Audience Research', in *European Journal of Communication*, vol. 2 no. 1, 1987.
3. J. Fiske, J. Hartley, *Reading Television* (London: Methuen, 1978), p. 85.
4. See T. Gitlin, *Inside Prime Time* (New York: Pantheon, 1983), p. 12.
5. V. Turner, 'Process, System and Symbol: A New Anthropological Synthesis', in *Daedalus*, vol. 106 no. 3, Summer 1977, p. 69. I am grateful to Daniel Dayan for introducing me to Turner's work.
6. Ibid., p. 68.
7. Ibid., p. 73.
8. Ibid.
9. T. Gitlin, 'Prime Time Ideology: The Hegemonic Process in Television Entertainment', in *Social Problems*, vol. 26 no. 3, 1979.
10. H. Newcomb, 'On the Dialogic Aspects of Mass Communication', in *Critical Studies in Mass Communication*, vol. 1 no. 1, 1984.
11. M. Bakhtin, *The Dialogic Imagination* (Austin: Univ. Texas Press, 1981), p. 291; quoted in Newcomb, ibid., p. 39. Cf. also S. Fish, *Is There a Text in This Class? The Authority of Interpretive Communities* (Cambridge: Harvard University Press, 1981).
12. The Boston Tea Party example corresponds closely to Fish's case in which students were asked to interpret a '17th Century poem' made up of six authors' names. See S. Fish, *Is There a Text in This Class?*
13. See D. McQuail, *Mass Communications Theory* (London: Sage, 1983).
14. Ibid., p. 155.
15. Ibid., p. 156.
16. Newcomb, p. 36.
17. In this article I quote from four of the 16 interviews conducted: no. 3: AB, a 40-year-old woman – a married office worker with two children and no college education; no. 4: SC, a 20-year-old woman – a married assistant cook with no college education; no. 7: GO, a 65-year-old woman – a retired hospital catering officer, married, with some college education; no. 15: JH, a 55-year-old male traffic warden, married, with one child.
18. For the origins of these insights see N. A. Nielsen, *Seriesening i tv* (Danish Broadcasting Research Report, 1982), and E. Katz et al., 'On Commuting Between Television Fiction and Real Life', unpublished paper, Annenberg School of Communications, University of Southern California, Los Angeles, 1983.
19. McQuail, *Mass Communications Theory*, p. 160.
20. Ibid.
21. E. Katz, T. Liebes, 'Once Upon a Time in Dallas', in *Intermedia*, vol. 12 no. 3, 1984; T. Liebes, 'Ethnocriticism: Israelis of Moroccan Ethnicity Negotiate the Meaning of *Dallas*', in *Studies in Visual Communication*, vol. 10 no. 3, 1984; J. Radway, 'Women Read the Romance: The Interaction of Text and Context', in *Feminist Studies*, vol. 9, 1983; J. Radway, *Reading the Romance: Women, Patriarchy and Popular Culture* (Chapel Hill: Univ. North Carolina Press, 1984).
22. J. Fiske, J. Hartley, *Reading Television*, p. 80.
23. Ibid.
24. Thorburn, 'Television Melodrama', p. 530.
25. Ibid., p. 532.

26. J. Williamson, *Decoding Advertisements: Ideology and Meaning in Advertising* (London: Marion Boyars, 1978), p. 76.
27. Ibid., p. 72.
28. Ibid., p. 73.

SONIA M. LIVINGSTONE

Viewers' Interpretations of Soap Opera: The Role of Gender, Power and Morality

Three Types of People Who 'Read' Television Programmes

Of all the people who enjoy, discuss, monitor, and for different reasons, watch and variously interpret in various ways television programmes, I am concerned with three categories of people: the academic psychologist, researchers in cultural studies, and the general public. The issue I wish to address is that of how television contributes to the social reality of the viewer. I shall approach this issue by discussing the differences between and problems within the disciplines of social psychology and cultural studies insofar as they deal with the mass media, and suggest some potential areas of integration, both theoretical and empirical, for research on the relation between television programmes and the viewers' social understanding.

Academic psychologists' 'readings' of programmes are largely implicit, detectable only from close examination of the assumptions underlying their experiments on the effects of television. Psychologists rarely spend time analysing or even describing the programmes or genres with whose effects they are concerned, partly because of their largely implicit behaviourist semantics. From references to 'incorrect comprehension', 'biased recall', and 'developmental deficits', one can detect the assumption that programmes, indeed social stimuli in general, are so transparent, unambiguous, and unitary in meaning as to require no analysis. In other words, the transmission of a programme is held to lay bare a simple and obvious meaning such that viewers can be judged right or wrong in their interpretations. While the dual assumptions of a passive viewer and a transparent text have earned psychologists some scorn from other disciplines, they have also caused problems within psychological research, and have contributed to the general disillusion with the 'effects' enterprise.[1] Indeed, if programmes cannot be recognised as structured, communicative products, then their communication with viewers will be misconstrued. It seems fair to say that psychologists tend to study the closed and denotational aspects of programmes and miss those more open, rhetorical or connotational aspects with which the viewer is most involved and where their own opinions and knowledge have the greatest role to play. In other words, psychologists often study a different programme from that experienced by the viewers, for psychologists and viewers make different readings.

Rather than steaming ahead with 'effects' research on an ever-larger scale,[2] I believe we should slow down and think again. Let us now take seriously what is taken for granted in other domains. Firstly, cognitive and socio-cognitive psychology take for granted the constructive and schematic nature of perception and representation, as well as the concept of levels of preconscious and conscious processing.[3] While the latter point complicates our research methodologies, the former challenges all notions of uptake, miscomprehension or inaccurate perception, and representation devoid of other social knowledge. So, within mainstream psychology there exists a much more complex and active conception of the perceiving subject, standing in a different relation to the 'stimulus'. However, the 'stimulus' has not been reconceptualised in complementary fashion.

Secondly, literary criticism, semiotics and cultural studies take for granted the complexity of textual semantics, with their concepts of unlimited semiosis, rhetoric, closure, levels of meaning, presupposition, and communicative aim.[4] This stands in direct contrast to the 'text as stimulus' view of psychology.

The first point is gradually being recognised by psychologists studying the media. In interpreting a programme, it is argued that viewers draw upon the information presented by the programme, their past experiences with the programme and its genre, and their own personal experiences with the social phenomena (institutions, relationships, personalities, and explanations) referred to by the programme. So, before television can have behavioural, attitudinal or cognitive effects, a programme must first be perceived and comprehended by viewers. To the extent that viewers' representations of a programme are a distortion, selection, or transformation of the original, the programme's consequences will also be affected. However, the latter point, concerning textual semantics, still is largely ignored by psychologists. Those media researchers who recognise textual complexity tend not to be those concerned with the psychology of the viewer or, with some exceptions, with the interface between the two. Yet for the study of television's effects on viewers, we need to be conceptually clear about exactly what we think may effect what, and with what causal power.

The second set of programme readers then are those in the area of cultural studies, semiotics of the media, etc. They apply the analytic tools gained from the study of 'high' culture, namely literature, to that of popular culture.[5] Such analyses are often illuminating about both the programmes themselves and, for Marxist-oriented researchers, the culture and organisation which produced them.[6]

More recently, however, this area has also come to recognise certain inherent problems in its aims and assumptions. Specifically, the attempt to discover the true, hidden meaning of texts has led to: a crystallisation or reification of the text; a neglect of the interface with the reader; a failure to discover a touchstone of truth, that is, a relativism in text readings by different researchers; and last but not least, a wealth of impenetrable jargon. The concern now is to determine the ways in which texts mean, engage or anticipate their reader, open up or close down possibilities, and so forth. Thus any analyses of television programmes must embrace rather than ignore the interpretative role of the viewer, for this is anticipated in the creation of programmes.

Often, recent writers in the two domains use similar concepts – for example the Gestalt schemata used by both reception theorists and the social cognition school – or have similar aims, for example to study the realised text, namely that which results from the interaction of text and reader, instead of the virtual text (of interest to structuralists) or the effects on the reader (of interest to psychologists).[7] This shift in focus, from the meaning of the text to the intelligibility of the text parallels a shift in focus within psychology from the effects of television towards how television can have effects: in all, a shift from asking what to asking how.

Ultimately, of course, the viewers are the focus of all this theorising. Their readings, the choices or interpretations which they make and the importance of television programmes to them have been very much neglected, although there are some notable exceptions to the neglect of viewers' experience of television.[8] Literary critics generally, and often those in cultural studies, argue against the empirical investigation of the role of the reader or viewer in contributing to the meaning of the text, despite their own arguments which suggest text analysis is incomplete without this. They argue that such investigation is unrewarding: the results can easily be anticipated and often contribute little to the understanding of literary texts. While these arguments are justly motivated by the desire to avoid the psychological reductionism or individualism of experimentation, there are social psychologists engaged in empirical research with similar motivations.[9] Moreover, neither of the above excuses is acceptable; even foreseeable results need demonstration, and furthermore, what is foreseeable is often only so in retrospect. Results are harder to foresee when the readers are the general public, as they are for the media, and not one's students or colleagues, as when reading literature. But more importantly, as a social psychologist, my concern is not so much with the role viewers may play in illuminating the meaning of texts, but instead is with the role that television texts may play in illuminating the symbolic world of the viewers.

The Research Domain

My aim, then, is to conduct research into viewers' interpretations of television programmes which goes some way towards effecting an integration between the above two domains of media research, psychology and cultural studies. This research draws upon those theories and assumptions of the two domains which I have indicated above show certain compatibilities or overlap in terms of the interaction between text and viewer, and it should provide results which make sense to both domains. Naturally, any research can only deal with one small aspect of a problem, and I shall briefly justify two restrictions I have placed upon my research. The first is that I have only investigated interpretations of soap opera, in particular *Dallas* and *Coronation Street*.

Soap opera is a genre of social realism and therefore it explicitly aims to parallel or directly contribute to the symbolic world of the viewer. It comprises relatively open texts, texts which maximise the interpretative role of the viewer by inviting application of the viewers' personal and social knowledge. It presents a wide range of characters and issues; particularly, it includes as many females as male characters, and it is viewed for a wide range of motivations.[10] Finally, it is, of course, one of the most popular genres of television programming.

Secondly, I selected one major aspect of soap opera, that of characterisation. This was because this has aroused the interest of various media researchers, because it relates clearly to various psychological theories (for example, person perception, stereotyping), and because it is a prerequisite for various important theories of media effect (for example, role-modelling, identification and para-social interaction).[11]

What Can an Investigation of Character Representation in Soap Opera Tell Us?

The first and most simple justification is that of describing a very common experience: how do viewers experience soap opera characters and according to what basic themes? However, having such a description, together with methods for obtaining such descriptions, is of value in relation to the proposals and problems of both social psychological and cultural studies research.

Let me give some examples of typical psychological research, and indicate how a knowledge of viewers' interpretations of characters could be valuable. Firstly, consider the correlational research exemplified by Gerbner and his co-workers. Some of this research finds significant correlations between, say, stereotyped attitudes and amount of television viewing, and some does not.[12] This research has been refined by correlating endorsements of particular attitudes with the frequency of viewing programmes believed to encourage those attitudes,[13] but still with mixed results. The problem lies in knowing what to correlate with what. I suggest that the programme themes which viewers find most salient, or around which they construct their own interpretations, might be the locus for any attitude/viewing correlations.

Secondly, researchers have attempted to demonstrate children's comprehension of, and therefore potential influence by, programmes by examining whether they can correctly or incorrectly recall the narrative. The potential for such paradigms[14] might be extended if one knew which themes viewers considered irrelevant and which they used as the basis for organising their understanding, as it is in relation to the latter that one would expect to find most 'effects'.

Both of these examples are concerned with cognitive effects, and both place increased emphasis upon the role of the viewer in actively making sense of television programmes. Both examples also emphasise the importance of relevance; as Grice has made clear, in a communicative process, the psychological concept of relevance is central in clarifying which, of a set of possibilities, was in fact selected.

My interest is in how to study the ways in which, and the extent to which, the media reinforce or alter people's cognitive world view, to study the media's contribution to the symbolic construction of reality. Other researchers have been more concerned with the role of the media in relation to behavioural or personality issues, specifically research on role-modelling, imitation and identification. This research is concerned precisely with television characters, but the mediational role of interpretation and representation of these characters has received little attention.

While there has been a very just concern with the distribution of television role models – few women, violent men, incompetent elderly people – the cognitive representational aspect is also important: of the characters presented, viewers select particular models, and their criteria, interpretations, and constructions are important mediating factors additional to the importance of the set agenda provided.

Finally, for research on parasocial interaction – the way in which people appear to treat television characters as they do real-life acquaintances, the way in which people interpret the characters and the relation of this to their interpretation of real-life people – is an open issue.

Let me turn now to the viewpoint of semiotics and cultural studies, with its increasing interest in psychological issues. This research domain makes two types of claim regarding the ways in which viewers interpret or represent television characters: first, that viewers identify in programmes the same themes or basic codes which textual analyses suggest they must; second, that these themes fit the formal structure frequently proposed by such researchers, namely that of binary oppositions.

Thus, texts are frequently analysed by identifying, more or less loosely, the basic oppositions around which the text is structured, thereby going beyond the manifest level typically of concern to psychologists to the deeper, connotative meanings. The concept of oppositions is often incorporated within a more complex semiotic theory of the text, and therefore of text reading, whereby the oppositions describe the paradigmatic, rather than the syntagmatic aspect of meaning, namely the similarities and contrasts created at any moment in the narrative.[15] The openness in the text, Allen argues, lies in the paradigmatic aspect of the programme, in the meanings created by the contrast with what might have happened, with the alternatives which were possible but not selected. The text of a soap opera cannot be identified as any particular episode, but instead consists of the sum total of all previous episodes. This is because the paradigmatic aspect is established through the characters' interrelations over time and is often present not 'in' the text itself so much as in the text's reference to its own past, to the viewer's memory. Thus, perceptions are framed and informed by past conflict, contrasts, judgments, and actions – gossip, advice, sarcasm, gain their meaning by reference both backwards in the history of the programme as well as by reference outside the programme to viewers' more general wealth of social and moral knowledge. If one accepts such an analysis, and I think it has much to offer, then an investigation of whether binary oppositions underlie viewers' representations of characters, and of what those oppositions are, will also give us a handle on plot development: scene-setting, disruption, tension and resolution may be understood in part as the playing out of the various central oppositions of the text.

I have now introduced the notion of testing out cultural studies' notions of character representation in terms of the oppositions that viewers 'should' identify as central if the text is to be made meaningful. Let me now reintroduce the psychology versus cultural studies theme and claim that psychological theories about the basic themes which underlie person, and presumably, character perception can also be tested in relation to television characters. And to pursue the notion of oppositions, it is convenient that the main theories of person perception also propose bipolar dimensions as the interpretative tools for making sense of the complex texts presented, whether they be television or the social world. Thus, of all the possible ways in which links might have been created between psychology and cultural studies (for example, via story grammar research, or schema theory), I have capitalised on the parallel concepts of oppositions and dimensions in the two domains to pursue one possible area of integration.

Empirical Research on Character Representation

Turning now to my empirical research, I shall report two studies of character representation, one of *Dallas*[16] and the other of *Coronation Street*. These programmes were selected as prototypical American and British soap operas and they differ in many dimensions. The research falls into two phases.

The first phase asks whether viewers have a stable and consensual representation of the characters in these soap operas. Multidimensional scaling[17] was selected as the appropriate method because an exploratory rather than a hypothesis-testing procedure was required initially to determine which underlying dimensions of character representation emerge spontaneously from subjects' judgments. The task requires subjects to assess the characters in terms of their similarity of personality, a deliberately general criterion, and the analysis makes explicit the criteria implicitly underlying their judgments. This is achieved by generating a spatial model which positions the characters according to the most explanatory or basic dimensions which underlie judgment. Conceptual similarity or dissimilarity between characters is represented by proximity or distance in multidimensional space. The advantage of this method is that subjects' judgments are not affected or constrained either by the theoretical assumptions of the researcher or by the task of making their underlying criteria explicit.

The second phase involves the interpretation of the space. This is achieved by operationalising various theoretical predictions of the oppositions or dimensions. Rated attributes of the characters are fitted statistically on to the character representation: those which do not fit, or only fit poorly, are considered irrelevant to the viewers' representation, while those which fit closely are taken to be those themes, or semantically close to those themes, which implicitly underlie the viewers' character representation.

Phase 1 – The Character Representation for 'Dallas' and for 'Coronation Street'

Method for 'Dallas'

CHARACTERS The central characters in *Dallas* at the time (December to February 1984–5) were as follows: J.R. Ewing, Bobby Ewing,

Pam Ewing, Jenna Wade, Sue Ellen Ewing, Miss Ellie Ewing, Ray Krebbs, Donna Krebbs, Lucy Cooper-Ewing, Cliff Barnes, Katherine Wentworth, Clayton Farlow, and Sly (surname unknown – J.R.'s secretary).

SUBJECTS   Subjects were 40 *Dallas* viewers. They were 23 females and 17 males drawn from a wide range of occupations and ages. 27 watched *Dallas* 'regularly', with 7 viewing 'fairly often', and 6 viewing 'sometimes'. All subjects knew all characters except for two who did not know Sly. 27 subjects were British, from the University of Oxford's Department of Experimental Psychology subject panel and were paid for participating. The remaining 13 subjects were Canadians who volunteered through a Canadian colleague.

QUESTIONNAIRE AND PROCEDURE   The 78 possible character pairs were used to make two alternative forms of the questionnaire. These were matched so that when in one form, a pair was presented in the order AB, it was presented as BA in the other. A single random order of the 78 pairs was generated which occupied three pages, the order of which was counterbalanced across subjects. For each pair, the similarity of one pair member to the other was rated by subjects using a five-point rating scale defined by the expressions 'not at all similar' and 'very similar'.

Method for 'Coronation Street'

CHARACTERS   Twenty-one characters in *Coronation Street*, 11 women and 10 men, were considered to be both longstanding and central to the programme: Ken Barlow, Deirdre Barlow, Vera Duckworth, Jack Duckworth, Terry Duckworth, Kevin Webster, Mike Baldwin, Curly Watts, Sally Waterman, Ivy Tilsley, Brian Tilsley, Gail Tilsley, Billy Walker, Emily Bishop, Mavis Riley, Rita Fairclough, Bet Lynch, Betty Turpin, Percy Sugden, Hilda Ogden, and Alf Roberts.

SUBJECTS FOR 'CORONATION STREET'   The 58 subjects were obtained through the Department of Experimental Psychology subject panel and were paid for participating. There were 47 women and 11 men. This imbalance was due to the difficulty of obtaining men who viewed the programme, despite efforts made to obtain male subjects. Otherwise, efforts were made to obtain a varied group of regular viewers, so as to approximate the actual viewing population. Twenty-two subjects were under 30 years old, 30 were between 31 and 60, and 6 were over 60. There were 14 homemakers, 18 white collar workers, 9 professionals, 4 unskilled workers, 2 unemployed and 11 students. Fourteen subjects had watched *Coronation Street* for between 2 and 5 years, 20 had watched between 6 and 10 years, 6 had watched for between 10 and 20 years, and 18 had watched for longer than 20 years. Thirty-one subjects claimed to watch the programme twice a week (i.e., every episode), 22 watched once or twice a week, and 5 watched less than once a week. Finally, all subjects knew all the characters with the exception of 6 subjects who did not know Sally Waterman, and one who did not know Billy Walker, Percy Sugden, and Kevin Webster.

PROCEDURE   The character names were typed on to separate slips of paper. Each subject received the 21 slips of paper, several paper clips, and the following instructions in an envelope:

> Each of the slips of paper in this envelope has the name of a *Coronation Street* character typed on it. You are asked to sort the 21 characters into piles according to how they *appear to you* in the programme. You may make as many piles as you want, and you can put any number of characters in each pile. If you don't think a particular character 'goes with' any other (or if you don't know who a character is), then put it on its own. Characters put into the same pile should be perceived by you as having *similar personalities* to each other, and as having different personalities from the characters you have put into other piles. Feel free to rearrange your piles until you are satisfied. Then use the paper clips provided to fix the slips of paper in each pile firmly together and *replace in the envelope* along with this instruction sheet.

Results for 'Dallas'

The raw data were averaged over subjects and the mean similarity score for each pair-wise comparison of characters was entered into MRSCAL,[18] a classical multidimensional scaling programme for metric data. Over all the pairwise comparisons, the standard deviations varied from 0.423 to 1.248, indicating that the mean similarity scores were reasonably representative of the original data. Multidimensional scaling is a mathematical procedure for modelling the perceived similarities between a set of objects by representing them as distances in multidimensional space. Using an iterative procedure, the measured similarities between objects are progressively modified to fit into euclidean space, minimising any discrepancies between perceived similarities and modelled distances (as indexed by a measure of badness of fit or stress). In the resultant space, objects placed close together were perceived as similar by subjects and vice versa.

The coefficients of alienation (the measure of stress for MRSCAL) for solutions for all dimensions from 5 to 1 were: 0.063, 0.077, 0.100, 0.136, 0.207. The 'elbow' lies at two dimensions, with a coefficient of alienation of 0.136. As a result of this analysis, the two-dimensional solution was selected for interpretation. It can be seen in Figure 1.

91

The MRSCAL programme for multidimensional scaling was again
used, this time entering the number of subjects who placed each pair
of characters together as the proximity data for the analysis. The
resultant coefficients of alienation indicated that the three-
dimensional solution best represented the original data. This had a
coefficient of alienation of 0.142, which is satisfactory. The space can
be seen in Figures 2 and 3. These each represent one of the three
2-dimensional faces of the actual 3-dimensional cube produced by the
analysis. Thus they should each be 'read' in conjunction with the
other.

Phase 2 – Fitting Properties onto Character Representations

The above studies set out to discover viewers' implicit representation
of the characters in *Dallas* and *Coronation Street*. Converging operations point to a robust and generalisable three-dimensional solution.
The low stress obtained for both spaces indicates a high degree of
consensus among subjects. Replication of the *Dallas* space corrobo-
rates the high degree of consensus in subjects' implicit character
representation. Multidimensional scaling methods have uncovered a
stable, consensual representation of television characters. The solutions can be interpreted in either or both of two ways, namely intuitive labelling of the dimensions by the researcher or the projection of

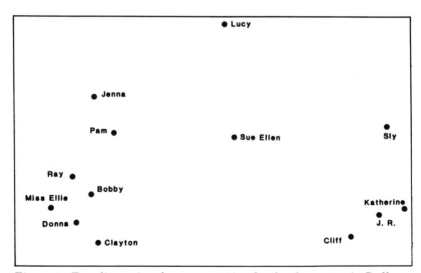

Figure 1 Two dimensional representation for the characters in *Dallas*
(dimension 1 = horizontal, dimension 2 = vertical)

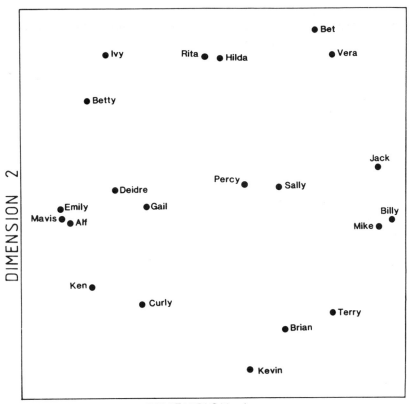

DIMENSION 1

Figure 2 Three dimensional representation for the characters in *Coronation Street* (dimensions 1 and 2)

property ratings for the set of stimuli onto the solution to determine which of a set of properties actually fit the space.[19] The second method overcomes the subjectivity of the first method, but interpretation is restricted to the particular set of properties measured, normally those of *a priori* theoretical interest. To maximise both the flexibility of the interpretation and the explanatory or hypothesis-testing potential of the emergent space, both methods will be used.

Any selection of properties to be tested as dimensions for the space must be limited on practical grounds. For the two programmes, the following dimensions were selected as being representative of the theoretical sources which 'should' explain the space. Implicit Personality Theory and Gender Stereotyping Theories are both social psychological theories of person perception which would be expected

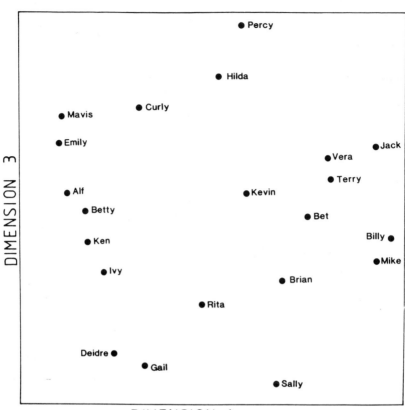

Figure 3 Three dimensional representation for the characters in *Coronation Street* (dimensions 1 and 3)

to predict the major dimensions underlying people's cognitive representation of television characters in addition to that of real-life people.[20] The oppositions listed in Table I as predictions from cultural studies are culled from various sources. Morality in soap opera in general is suggested by general research on soap opera as outlined above; the oppositions for *Dallas* and for *Coronation Street* are suggested by Mander and Dyer *et al.* respectively.[21] The remaining predictions were based upon an interpretation of the emergent spaces.

Method

QUESTIONNAIRE AND PROCEDURE   The ratings were presented to subjects in the form of a grid consisting of 13/21 columns, labelled with the names of the characters, and with 14/18 rows, labelled with

TABLE I
Predictions from Cultural Studies

Source	'Dallas'	Both	'Coronation Street'
Implicit Personality Theory		Intelligent/Unintelligent Rational/Irrational Sociable/Unsociable Warm/Cold Active/Passive Excitable/Calm Hard/Soft Dominant/Submissive	
Gender Stereotyping Theories		Masculine/Not masculine Feminine/Not feminine	
Cultural Studies		Moral/Immoral	
	Does/not value Family		Central/Peripheral to Community
	Does/not value		Middle/Working-Class
	Organisational Power		
Interpretation of Spaces	Pleasure/Business-oriented		Sexy/Not sexy
			Approach to Life (Modern/Traditional) Mature/Immature Roguish/Not roguish Staid/Not staid

the rating scales, as defined above. For each rating scale, one pole corresponded to a '1' and the other corresponded to a '7'. The following instructions were given to subjects:

> This questionnaire asks about *your views* of the characters in *Dallas/Coronation Street*. Over the page you will see a grid which has the names of the main *Dallas/Coronation Street* characters across the top and a list of personality traits down the side. You are asked to describe each of the characters in terms of each of the personality traits. Use the scale from 1 to 7 at the top of the page to make your judgment and base your judgment on your impression of each character in general as they appear in the programme.

An example was given of how to fill out the grid.

SUBJECTS FOR 'DALLAS'   The 20 subjects were all viewers of *Dallas* who volunteered to complete the questionnaire. Most of the 13 women and 7 men were students. Most (11) had been viewing *Dallas* for three years or more, 5 for two years, 3 for about one year, and one did not answer. Regarding the frequency of viewing, 7 viewed 'regularly', 6 'fairly often', 5 'sometimes', and two subjects omitted to answer. The subjects knew all of the characters except for four who did not know Sly or Katherine.

SUBJECTS FOR 'CORONATION STREET' The 21 subjects were obtained through the Department of Experimental Psychology subject panel and were paid for participating. There were 15 women and 6 men. This imbalance was due to the difficulty of obtaining men who viewed the programme, despite efforts made to obtain male subjects. Otherwise, efforts were made to obtain a varied group of regular viewers, so as to approximate the actual viewing population. Fourteen subjects were under 30 years old, and 7 were between 31 and 60. There was one homemaker, 6 white-collar workers, 5 professionals, 5 unemployed people and 4 students. Six subjects had watched *Coronation Street* for between 2 and 5 years, 8 had watched between 6 and 10 years, 4 had watched for between 10 and 20 years, and 3 had watched for longer than 20 years. Five subjects claimed to watch the programme twice a week (that is, every episode), 14 watched once or twice a week, and 2 watched less than once a week. Finally, all subjects knew all the characters except 2 subjects who did not know Sally Waterman, and one who did not know Billy Walker.

Results

The mean property ratings were calculated for each character by collapsing over subjects. This data was then entered into the PROFIT programme,[22] together with each of two stimulus configurations. The configurations used were those of the first study in (a) three dimensions and (b) two dimensions. This made it possible to identify the interpretative gains made by the addition of the third dimension.

PROFIT is a method of fitting externally rated properties for a set of stimuli to a MDS space for those same stimuli through the use of multiple linear regression.[23] If the space was constructed through an implicit method, such as similarity judgments or a sorting task, then PROFIT can be used to relate the implicit or emergent dimensions to the explicit or theoretically derived properties. It outputs each property in the form of a vector projected on to the space whose direction describes the direction in which the property increases through the space.

Tables II and III show the degree of fit and significance between the rated properties and the character space. For *Coronation Street*, the R-squared values for the properties projected on to the first two, and onto all three, dimensions of the space show the gain in information by including the third dimension. Kruskal and Wish recommend interpretation of only those properties whose R-squared is signifi-

TABLE II

*Property ratings projected on to the implicit
character space for 'Dallas'*

Property	R^2 (2 *dimensions*)
Sociable	0.384
Warm	0.769***
Intelligent	0.439
Rational	0.406
Dominant	0.779***
Hard	0.807***
Active	0.511*
Excitable	0.219
Masculine	0.312
Feminine	0.361
Moral	0.876***
Family values	0.554*
Organisational power	0.708**
Pleasure/Business	0.550*

*p<0.05   **p<0.01   ***p<0.001

TABLE III

*Property ratings projected onto the implicit character space for
'Coronation Street'*

Property	R^2 (2 *dimensions*)	R^2 (3 *dimensions*)
Intelligent	0.269	0.276
Rational	0.213	0.273
Sociable	0.407**	0.479**
Warm	0.436**	0.437*
Active	0.157	0.125
Excitable	0.188	0.262
Hard	0.706***	0.804***
Dominant	0.485**	0.582**
Feminine	0.548***	0.664***
Masculine	0.724***	0.698***
Central to community	0.276	0.375*
Working-class	0.351*	0.314
Modern approach	0.370*	0.740***
Mature	0.408**	0.463*
Sexy	0.140	0.628***
Roguish	0.823***	0.888***
Staid	0.500***	0.721***
Moral	0.755***	0.850***

*p<0.05   **p<0.01   ***p<0.001

cant beyond the 0.01 level.[24] The character representations with the significantly fitting properties are shown in Figures 4–6.

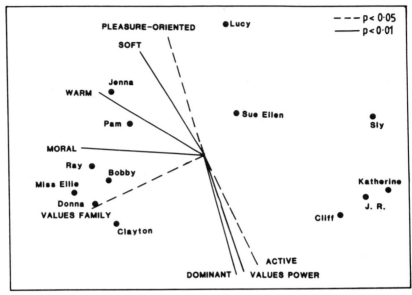

Figure 4   Two dimensional representation for the characters in *Dallas* (dimension 1 = horizontal, dimension 2 = vertical), showing projections of significant properties

Discussion

Relevant and Irrelevant Oppositions

From the emergent character representations and the fitted properties, a number of conclusions can be drawn about both the nature of viewers' interpretations of the programmes and about the theories which aim to account for these interpretations. The spatial representations show two things: the relations between the rated attributes and the characters, and the inter-relations between the attributes themselves.

Person perception theories suggest variants on a basic scheme in which evaluation (social and intellectual), potency, and activity comprise the basic, orthogonal dimensions which organise people's representations of others.

In *Dallas*, activity (active/passive and excitable/calm) was only weakly relevant; in *Coronation Street*, it did not appear at all in the space. In both programmes, intellectual evaluation (intelligent/

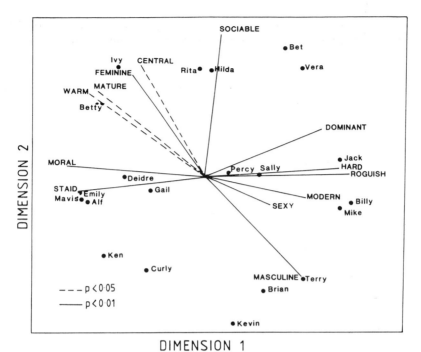

Figure 5 Three dimensional representation for the characters in
Coronation Street (dimensions 1 and 2), showing projections of
significant properties

unintelligent and rational/irrational) was irrelevant. Social evaluation (sociable/unsociable and warm/cold) was strong in *Coronation Street* and partially present for *Dallas*. In both programmes, potency (dominant/submissive and hard/soft) was very important as a basic theme of representation. As predicted, potency and evaluation were fairly orthogonal to each other.

In all, the predictions of Implicit Personality Theory were only partially supported. In relation to soap opera characters, viewers do not find relevant certain themes which psychologists have considered basic, but the concept of power is clearly relevant. In these respects, the results for the two programmes were very similar, suggesting both problems and strengths for the theory.

Gender stereotyping theories make two basic predictions. Firstly, in order to allow for androgynous interpretations (in which masculine and feminine qualities are compatible), masculinity and femininity should be orthogonal. Secondly, if masculinity and femininity are negatively correlated, as suggested by earlier stereotyping

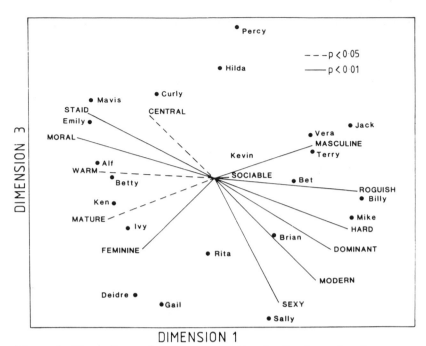

Figure 6 Three dimensional representation for the characters in *Coronation Street* (dimensions 1 and 3), showing projections of significant properties

theories, then masculinity should be semantically related to concepts such as power, activity, rationality, intelligence, while femininity should be related to their opposites.

The findings for the two programmes were very different. For *Dallas*, although more women than men occupy the weaker and softer half of the space, viewers do not perceive the characters in terms of the attributes of masculinity and femininity. The programme thus is seen to present mainly stereotypic women and men, in line with much else on television, but it also includes a few counterstereotypic female characters. These characters do fit lower-level stereotypes: Miss Ellie as matriarch, Katherine as the wicked witch. These counterexamples were sufficient to prevent a fit for the gender attributes in the representation.

For *Coronation Street*, masculinity and femininity fit the space very well, and are highly negatively correlated. This supports the strong emphasis placed on gender by both cultural studies and social psychological theories, although these latter conceive of masculinity and femininity as orthogonal. Thus, viewers make a clear distinction between the male and female characters when thinking about their personalities. However, femininity in *Coronation Street* is not of the traditional variety which is related to irrationality, softness, or weakness, as social psychologists propose, but is more matriarchal in character. Femininity is related to maturity, warmth, centrality to the community and sociability, in contrast to the rather cold and childish masculinity of the male characters. This was anticipated by cultural studies noting *Coronation Street*'s ample supply of 'strong, middle-aged women characters which the conventions of soap opera seem to require [and who] must remain independent'. Such portrayals may carry subversive significance, in contrast, say, to those of *Dallas*: '*Coronation Street* offers its women viewers certain "structures of feeling" which are prevalent in our society, and which are only partially recognised in the normative patriarchal order'.[25] Interestingly, the female characters are not perceived as more powerful than the men for gender was orthogonal to potency.

Considering the predictions of literary/semiotic approaches, Mander's interpretation of *Dallas* is strongly supported. The major theme of morality emerged. From its relation to the other two properties that she proposed, it can be seen as a conflict between traditional family values and the power of organisations to destroy or undermine those values. Viewers are clearly sensitive to the opposition of good and bad, for the same dimension emerged strongly in *Coronation Street* as well. This connects with the interest of recent researchers in soap opera as a form of contemporary myth: characters are consensually perceived according to moral themes. These themes are not, however, those of individual quest or the vindication of truth, but instead concern the tangle of human relationships and their dynamics as they face everyday minor and major personal dilemmas. Notice that in *Dallas* morality and power are independent of each other, whereas in *Coronation Street* they are related such that good means weak and bad means strong. This suggests that the programme is presenting a rather negative view of morality, and the suggestion that dominance must be bad and submission good. In both programmes morality is unrelated to gender, so women are presented as neither more nor less moral than men, which rather detracts from the cosy warmth of the matriarchal theme (in *Dallas*, gender is instead related to power).

Also of importance is the position of characters as central or peripheral to the community, and the relationship of these positions to working- or middle-class status.[26] Centrality was only marginally significant and class was irrelevant. Viewers did not discriminate among characters implicitly in terms of their class. As with the other attributes which did not explain the representations, the absence of class as a relevant opposition can be explained in one of several ways.

One could reject the theory which proposes a particular opposition: Implicit Personality Theory is thus partly wrong, gender stereotyping theories are even more wrong, Mander is right about *Dallas*, and Dyer *et al.* are rather less perceptive about class than they are about gender. Or, one could say, particularly in relation to the readings made by cultural studies, that theorists and viewers simply have different knowledge and interests and therefore make divergent readings. A stronger claim argues that these attributes may indeed be central to viewers' interpretations of characters, and their absence in the representations indicates false consciousness on the part of the viewer. Of course, one could instead blame the methodology, for although the sorting task was fairly natural and implicit, the operationalisations of theories into rating scales was relatively crude, and the method as a whole cannot capture the complexity of Dyer *et al.*'s analysis. The question is whether it can capture its essence.

In relation to the analyses of cultural studies, the picture is also mixed. But there is, I believe, sufficient support for their insights into soap opera as interpreted by viewers. In all, the present data suggest that both social psychological theories and textual analyses are needed to understand viewers' interpretations of soap opera characters. For cultural studies, the results suggest that theorising should not proceed too far and presume too much in ignorance of the actual viewers. For social psychology I would say the opposite: important themes such as morality, sexuality, and family values are neglected by theorists overconcerned to reduce the complexity of everyday understanding to a few, fundamental dimensions.

Let me now summarise the results from the two programmes by translating statistical findings into the formal oppositional structure familiar to semioticians. Each set of concepts below is more or less independent of the other, and each is internally related.

(1) *Coronation Street*

MORAL	IMMORAL
Staid	Roguish
Soft	Hard
Submissive	Dominant

(2) *Coronation Street*

FEMININE	MASCULINE
Mature	Immature
Warm	Cold
(Central)	(Peripheral)
(Sociable)	(Unsociable)

(3) *Coronation Street*

MODERN	TRADITIONAL
Sexy	Not Sexy

(1) *Dallas*

MORAL	IMMORAL
Warm	Cold
For Family Values	Not For Family Values

(2) *Dallas*

FEMALE	MALE
Submissive	Dominant
Soft	Hard
Pleasure-oriented	Business-oriented
Doesn't Value Power	Values Power
Passive	Active

So, for *Dallas*, we have a representation which presents the contrast between a (mainly female) world of pleasure, weakness, and femininity and a (mainly male) world of organisational power and hard-headed business. The programme does not portray the former, hedonistic world as either more or less moral than the latter, busi-

103

ness world, and with the power split equally between the 'goodies' and the 'baddies', the fight between them will be equal and endless.

For *Coronation Street*, the representation centres on three distinct themes. Fortunately the stakes are not so high as in *Dallas*, for the first theme aligns morality and power so that the immoral characters are seen as more powerful than the moral. Secondly, the matriachal women are opposed to the more cold and peripheral men with no real exceptions although there is an area of overlap. This theme is unrelated to morality and power, again in contrast to *Dallas*, so the roles of men and women are equal in these respects. The final theme is that of approach to life and sexuality, and suggests a conflict between traditional, even nostalgic, domestic stability in the face of the exciting and new challenges of the future.

The Status of the Character Representation
There are several points to be made about the status of the representations produced by the present research:

(a) These representations might be regarded as forming a literal 'picture in the head'. Easy as it is, however, to imagine people 'having' character representations which they carry about in their heads and match up to television characters, noting discrepancies and so forth, such a view is unacceptable for several reasons. It is psychologically reductionist. It reifies knowledge, creating the problem of its relation to interpretation. It brings with it all the limitations of semantic memory theories in cognitive psychology. It is also philosophically incoherent.

(b) One might see the representations as providing a model of interpretative processes. The representation can be regarded as a researcher's schematic summary of viewers' interpretative response to the characters in the programme. Of the many diverse resources and strategies with which viewers approach a programme, and of the many themes, narratives, and judgments which the programme presents to viewers, this is a summary representation of the emergent product of their interaction over time. This is a more dynamic approach with more theoretical potential. It is still, however, open to the charge of psychological reductionism, allowing for a very private viewer/programme interaction.

(c) One might think in terms of a 'Social Representation'. This concept, taken up and developed recently by Moscovici and colleagues,[27] is an attempt to further 'socialise' social psychology by integrating psychological processes of perceiving and interpreting social phenomena with the social context of knowledge, prejudices, consensual images and myths, political atmosphere etc. within which individuals are constituted. As an idea, it is most in common with (b), although it is also compatible with (a): the representations could be taken as cultural pictures in the group mind, generally available representations of social origin of which individuals each 'have' a version. But the combination of (b) and (c) is the most interesting, and avoids the problem of reification. I suggest that the character representations obtained in the present research should be regarded as a researcher's 'handle' upon the consensual, conventional, historically-contexted interpretations and interpretative processes which relate to, in this case, soap opera characters. Neither 'in' the programmes nor simply the knowledge and biases of people ignorant of television, these interpretations emerge from the communication between the two. This more dynamic view of character representations, which emphasises relations between rather than features of the characters, relates also to the understanding of plot. One can describe a structured set of interrelations between the characters in the representation within which the character movements through a plot may be located. Any particular plot may be understood in terms of this character space by describing the movement of characters along the dimensions or oppositions of the space, with the consequent changes in character positions setting up narrative tension for the re-establishment of the original arrangement.

Concluding Remarks

Comment should be made on the methodology adopted to gain a 'handle' on viewers' interpretations. I do not wish to claim that this is the only way interpretations could be represented, or that these representations capture every aspect of interpretations. My strategy was to see how far we can go by representing interpretations as spatial models with bipolar dimensions. However, if the methodology was truly inappropriate, the space could not statistically have been produced, and no attributes could have been fitted to it. The success of the method therefore justifies these initial assumptions. It also opens the way for further research on viewers' interpretations of television, whether this is research on different programmes or genres, or on the comparison between the representations of different categories of viewer, or on the testing of predictions from different theories.

Throughout this essay I have deliberately fudged the issue of the determination of the character representations. Clearly representations derive from both the structure of the television text and the motives, schemata, and knowledge of the viewer, as well as from their mutual cultural background. Viewers' representations of characters must be an interactive product of all three: they are 'realised' texts of complex origin. Yet I cannot see how to separate these empirically if, with an interactionist perspective and a rejection of the ideas of independently existing texts, one can have no independent access to texts prior to being read or to viewers ignorant of television texts. The relative power of the text to impose upon or direct the viewer and the power of the viewer to interpret or select from or transcend the text is an important question to which there is no simple answer. We still need to find ways of studying the 'preferred' readings or hidden meanings of the text and the interpretative strategies of the viewer. In other words, we need to study the means by which both text and viewers attempt to exert power, before we can see who 'wins'. Are viewers more receptive or constructive? Are texts more indeterminate or directive? This is one research project which I see for the future. Yet uncertainty regarding the origins of the representations does not prevent study of their nature and consequence. If we treat them as Moscovici's social representations – existing social phenomena – then another question for psychologists is: how do the representations connect with and contribute to people's world view?

This research was conducted while the author was supported by a grant from the Economic and Social Research Council, UK in collaboration with the Independent Broadcasting Authority, UK. The author would like to thank Michael Argyle and Peter Lunt for their helpful comments and suggestions on earlier versions of this paper.

References

1. See, for example, D. F. Roberts, C. M. Bachen, 'Mass Communication Effects', in *Annual Review of Psychology*, vol. 32, 1981.
2. See E. Noelle-Neuman, 'The Effect of Media on Media Effects Research', in *Journal of Communication*, vol. 23 no. 3, 1983.
3. See D. J. Schneider et al., *Person Perception* (Wokingham: Addison-Wesley, 1979).
4. See J. Fiske, J. Hartley, *Reading Television* (London: Methuen, 1978).
5. See, for example, R. Dyer et al., *Coronation Street*, TV Monograph 13 (London: British Film Institute, 1981); H. Newcomb (ed.), *Television: The Critical View* (Oxford: Oxford University Press, 1982); L. Masterman (ed.), *Television Mythologies: Stars, Shows, Signs* (London: Comedia/MK Media Press, 1984).
6. See S. Hall et al., *Culture, Media, Language* (London: Hutchinson, 1980).
7. See U. Eco, *The Role of the Reader: Explorations in the Semiotics of Texts* (Bloomington: Indiana University Press, 1979).
8. See D. Morley, *The 'Nationwide' Audience: Structure and Decoding* (London: British Film Institute, 1980) and D. Hobson, *Crossroads: The Drama of a Soap Opera* (London: Methuen, 1981).
9. See, for example, S. Moscovici, 'The Phenomenon of Social Representations', in R. M. Fos, S. Moscovici (eds.), *Social Representations* (Cambridge: Cambridge University Press, 1984).
10. On soap opera and social realism, see R. Dyer et al., *Coronation Street*; on its 'open' texts, see R. C. Allen, *Speaking of Soap Operas* (Chapel Hill: University

of North Carolina Press, 1985); on motivations, see M. Cantor, S. Pingree, *The Soap Opera* (Beverly Hills: Sage, 1983).

11. On person perception, see Schneider et al., *Person Perception*, and on stereotyping see S. L. Bem, 'Gender Schema Theory: A Cognitive Account of Sextyping', in *Psychological Review*, vol. 88 no. 4, 1981, and H. Markus, 'Self-Schemata and Processing Information About the Self', in *Journal of Personality and Social Psychology*, vol. 35, 1977. On role-modelling see G. Tuchman et al., *Hearth and Home: The Images of Women in the Mass Media* (New York: Oxford University Press, 1978); on identification, see B. Eisentock, 'Sex-Role Difference in Children's Identification With Counterstereotypical Televised Portrayals', in *Sex Roles*, vol. 10, nn. 5–6, 1984; on para-social interaction, see G. Noble, *Children In Front of the Small Screen* (Beverly Hills: Sage, 1975).

12. See R. P. Hawkins, S. Pingree, 'Television's Influence on Social Reality', in E. Wartella, D. C. Whitney (eds.), *Mass Communication Review Yearbook*, vol. 4 (Beverly Hills: Sage, 1983).

13. See A. Alexander, 'Adolescents' Soap Opera Viewing and Relational Perceptions', in *Journal of Broadcasting and Electronic Media*, vol. 29 no. 3, 1985.

14. W. A. Collins et al., 'Age-Related Aspects of Comprehension and Inference From a Television Dramatic Narrative', in *Child Development*, vol. 49, 1978.

15. See Allen, *Speaking of Soap Operas*.

16. See S. M. Livingstone, 'The Implicit Representation of Characters in *Dallas*: A Multi-dimensional Scaling Approach', in *Human Communication Research*, vol. 13 no. 3, 1987.

17. See J. B. Kruskal, M. Wish, *Multidimensional Scaling* (Beverly Hills and London: Sage, 1978).

18. See E. E. Roskam, *The MDS(X) Series of Multidimensional Scaling Programmes* (Edinburgh: University of Edinburgh Programme Library Unit, 1981).

19. See Kruskal and Wish, *Multidimensional Scaling*.

20. On Implicit Personality Theory see Schneider et al., *Person Perception*, and R. D. Ashmore, F. K. Del Boca, 'Sex Stereotyping and Implicit Personality Theory: Toward a Cognitive-Social Psychological Conceptualisation', in *Sex Roles*, vol. 5 no. 2, 1979. For Gender Stereotyping Theories see Bem, 'Gender Schema Theory', and Markus, 'Self-Schemata and Processing Information'.

21. Dyer et al., *Coronation Street*, and M. Mander, '*Dallas*: The Mythology of Crime and the Moral Occult', in *Journal of Popular Culture*, vol. 17 no. 2, 1983.

22. See Roskam, *The MDS(X) Series*.

23. See J. J. Chang, D. J. Carroll, *How To Use PROFIT, A Computer Program For Property Fitting By Optimising Nonlinear or Linear Correlation* (Murray Hill, New Jersey: Bell Laboratories, 1968).

24. See Kruskal and Wish, *Multidimensional Scaling*.

25. T. Lovell, 'Ideology and *Coronation Street*', in Dyer et al., *Coronation Street*, p. 50.

26. See M. Jordin, 'Realism and Convention', in Dyer et al., *Coronation Street*.

27. See Moscovici, 'Phenomenon of Social Representations'.

ERIC MICHAELS

Hollywood Iconography: A Warlpiri Reading

Isolated Aboriginal Australians in the Central Desert region, where traditional language and culture have survived a traumatic hundred years' contact period, began to view Hollywood videotapes in the early 80s and are now beginning to receive television from the new national satellite, AUSSAT. This situation raises many questions for humanistic research, including questions about the ability of the traditional culture to survive this new electronic invasion. I spent three years living with Warlpiri Aborigines of the Yuendumu Community undergoing this imposed transition, partly engaged in applied research and development leading to the birth of an indigenous community television station which challenged government policy and licencing. In the process, I noted what many other field-workers in non-literate, third world and indigenous enclaves are recognising: that electronic media have proved remarkably attractive and accessible to such people where often, print and literacy have not.

We are used to selling literacy as a pro-social, pro-development medium. We are used to denigrating video and television as antisocial and repressive. In this paper I want to examine these biases and their sources in an effort to see if our thinking on these matters can't be brought more in line with indigenous people's own capabilities and preferences in order to lead to a perspective which might prove more helpful and less protectionist.

This is of theoretical interest as well, because we have come to regard the western world's media development sequence as somehow natural: from orality to literacy, print, film and now electronics. But Aborigines and other developing peoples do not conform to this sequence, and produce some very different media histories. For philosophers and historians such as Ong, Innes or even Lévi-Strauss who posit special equivalencies between oral and electronic society, these ethnographic cases ought to prove especially revealing.

The Fallacy of Unilineal Evolution of Culture
Despite our best efforts, anthropologists have been unable to undo the racist damage of the unilineal theory of cultural evolution developed by our discipline's founding fathers. The Victorian passion to

classify encouraged Edward Burnett Tylor and Lewis Henry Morgan, for example, to borrow from Darwinian biology a vulgar theory of evolution and apply it to the very different problem of cultural variation. This produced the idea that people too unlike the Victorian English represented less evolved stages in a single great chain of culture.

This thinking was then used to justify both a dismissal of the value of non-western cultures and the evangelical urge to civilise heathens and savages. The *reductio ad absurdum* of the fallacy was realised by the 1940s in Nazi cosmology. By then, anthropology had well and truly repudiated the theory of unilinealism, and produced evidence of the authenticity and distinctness of cultural types. The theory of cultural relativism which replaced unilinealism now seems naive, but its basic tenet – that cultures do not arise in a single historical sequence – remains an established canon of anthropological thought. This revision has not been conveyed to the popular mind, however, or even much to other disciplines. (The problem is a bit like Lamarckism in biology. Despite the evidence, and the theoretical incompatibility with accepted theory, the possibility of inheritance of acquired traits persists, even in biology textbooks which sometimes talk as though the giraffe got its long neck from stretching for leaves; certainly almost everybody else talks this way.)

Evidence of the persistence of unilinealism is everywhere, from popular thinking about 'primitives' (aided and abetted by the etymology of the very word) to third-world development agendas, to a remarkable variety of humanist scholarship. It matters not whether the value loading is Rousseauesque romanticism, or native advancement, or simple garden variety racialism. The point is that very few people believe what anthropology teaches: that indigenous, small-scale traditional societies are not earlier (or degenerate) versions of our own. They are rather differing solutions to historical circumstances and environmental particulars which testify to the breadth of human intellectual creativity and its capacity for symbolisation.

In the case of Aboriginal Australians, it may be said that comparative isolation from the network of world cultures for perhaps 40,000 years encouraged a continuous cultural sequence such as is to be found nowhere else. By contrast, modern European culture is assumed to have recent points of disjuncture (the Industrial Revolution; the World Wars; the electronic revolution), so that certain of our own cultural forms may prove only a few generations old. If Aborigines have had 40,000 years to elaborate a continuous cultural tradition, certainly there must be respects in which this degree of elaboration produces cultural results of extraordinary value. We can hardly justify a claim that Europeans proved more sophisticated than the Aborigines, unless we take the technology and will to subjugate as privileged measures of human values.

Unilinealism surely persists in our theories of media history. The idea that media are either signals or engines of cultural sequence is familiar at least since Innes and McLuhan. Both, in accounting for European media, simplify to a unilineal historical causality for the West. And both are guilty of generalising from here to some universal evolutionary sequence, using 'primitive' societies in curious and usually uninformed ways to illustrate their points. But the fault cannot be attributed to these scholars alone. The very most common terms we use to distinguish our civilisation from others – 'historical' from 'pre-historical' – are semantically loaded with precisely the same ammunition, and refer to writing as the pivotal event in this presumed sequence. It seems telling that when we searched for a new, non-pejorative term to describe the distinction which would contain contemporary non-writing cultures, as well as ancient ones in a class contrastive to our own, 'literate' and 'pre-literate' gained the widest currency. Attempts to unbias the matter by substituting 'non-literate' never really caught on. And the descriptor 'oral', itself mostly unexamined, became confused with 'tribal', a word whose technical meaning is inappropriate for many of the groups so classified.

All these terms underscore western society's perhaps judaeo-derived cultural emphasis on our own writing skills, and reveal our opinion that people who can't write are backward. This very deeply rooted conceptualisation is now being challenged as we move from writing to electronic coding as the central symbolising system of our age. Marshall McLuhan looked for precedents and claimed that this new age would put us more in common with 'pre-literate', 'tribal' societies. And whether the public read McLuhan or not, it began to panic, aware that something was happening to our reliance on reading. Illiteracy rates appeared to be rising among our own young. The very fabric of culture and society was being sundered and we risked degenerating into some new form of savagery.

One problem with this literate myopia has been that we haven't paid much attention to what other people do instead of writing, and how information is processed, stored, transmitted, shared or received in its absence. This paper offers a reconsideration of the sequence of media history in a society which doesn't write, or at least has been believed not to.

Aboriginal Australian Media
Contemporary Aboriginal Australians have generally resisted several generations of concerted literacy teaching, first from Bible-toting missionaries and then from an 'enlightened' school system. Where-

ever their traditional law, language and culture are viable in Australia, literacy rates remain well below twenty-five per cent, even when bilingual and special education services have been available. I believe this implies that there is something essential to cultural maintenance associated with *not* writing, which is yet to be understood. But in this same population, videorecorders have appeared in the last few years, and during my three years of fieldwork I was able to observe the rapid adoption of electronic media by people who rejected print. This provides us with the intriguing but perhaps no longer so unusual situation of a people's moving rapidly from 'oral' to electronic society, but bypassing print literacy. Attention to the particulars of both the traditional system, and the accommodation to the imposed one, offers insights into the limitations of our unexamined theories of unilineal media evolution.

Traditional Aboriginal society did not employ, and presumably did not discover, alphabetic writing. It is probably accurate to classify this culture as 'oral', and elsewhere, I have discussed at some length some of the nature and consequences of this central feature of the society.[1] Yet Aborigines were not without resources for codifying experience and inscribing it in various media.

The Warlpiri people of the Yuendumu Community with whom I worked are noted particularly for a rich graphic design tradition described to us at length by Nancy Munn.[2] This design system extends throughout the central region of Australia and across most of the western desert, although symbols and techniques vary in particulars, from community to community. It appears that the northern coastal communities employed an appreciably different design system, which may correspond partly with the classification of many groups in this area as comprising a language family distinct from most of the rest of the continent. Graphic systems in the southeast quadrant are apparently defunct, along with most of the languages, so that rock art and artefacts contain suggestive clues but probably are insufficient for reconstructing the system as a whole, compared with the opportunity to observe the system in full use, as among the Warlpiri. Thus, it will be difficult to generalise accurately from the Warlpiri example, but I suspect that the basic graphic system described for them will prove to apply fairly widely to many Aboriginal societies in general outline.

Warlpiri Graphics: a Traditional Medium
Warlpiri design is a form of inscription which operates quite differently from more familiar writings. It neither provides phonological symbols which combine to record speech (alphabets), or pictorial glyphs which denote specified objects or ideas (picto/ideographs). If anything, it poses a sort of mediation between the two. It does pro-

112

vide pictographic symbols which are recombined to express ideas and things, but unlike Oriental ideographic writing, and more like alphabets, it abstracts elements so combined and reduces them to comparatively few. Attempts to provide lexicons which associate the symbols with discrete words, or even broad semantic domains, prove unsuccessful and ultimately misleading, as Munn seems to have discovered.

The system relies on a fairly discrete but polysemous inventory of graphic symbols: circles, dots, lines, semi-circles (additional, representational depictions of a specific tree leaf or animal body can be included – attempts by reviewers such as Dubinskas and Traweek to treat Munn's corpus as complete and rule-governed are incorrect).[3] These are combined in groundpaintings (made of soils, ochres, feathers, flowers), body decoration (ochres, pipeclay, animal fats), utilitarian objects and weapons, and special ceremonial objects (boards, stones, sticks) as well as being painted and carved on caves and rock faces. In recent years, traditional designs have been translated to acrylics and applied to canvas, objects, and murals, and women have developed a batik technique. Most of my experience of Warlpiri graphics comes from assisting in the formative stages of this transition to modern media for sale to contemporary art markets.

The distinctive features of Warlpiri graphics are many; I want to consider some of these which contrast it to contemporary writing systems. This is not the approach taken by Munn, or more recently by her critics. The grammatical structures and social/psychological functions of the design system for the Warlpiri which is discussed by these authors is accepted here as evidence of the richness of that system which can support various interpretations. Instead, I want to ask, as Havelock might, 'What kind of writing system is this?'[4] And I will conclude that it is as close to what we call writing (logo/ideographic or alphabetic) as you can get without compromising the authority of human speakers and interpreters: that is, a writing in the service of orality. Havelock would call Warlpiri graphics solipsistic: they record ideas rather than speech, are evocative rather than denotative, and do not assure a given reading. Unfortunately, such systems do not attract his attention except to remark that others have been most careless in their classification and attentions to such systems, a point well taken.

Some differences between Warlpiri graphics and other writing systems are immediately visibly apparent. Both western alphabets and pictographic writing systems arrange symbols in linear sequence and are biased towards surfaces, such as scrolls and book pages, which frame and support the cumulative string so produced. For Warlpiri, the relationship between the symbol and the surface is more creative. The painter/writer arranges the designs in a manner that relates them to the shape of the object, body or ground contour on which they are applied. Perhaps this is why Warlpiri graphics get reviewed as art, not writing. But this attention to the inscription surface is appropriate because the contents which are typically inscribed are 'dreaming' stories which are the myths accounting for the land, its objects and people. This is like 'bricolage' in Claude Lévi-Strauss's sense, where components enlisted in the service of an abstract recombinative system still retain their associations with their origins.[5] Warlpiri graphics might be called a 'meta-bricolage' because the inscribing media, the messages, and the surfaces to be inscribed are all meaningful bricolage elements of a dynamic intertextuality.

Another way to describe these graphic constructions is as maps. As stories of the landscape they are also images of that landscape. To this extent, spatial relationships of the symbols are constrained by a spatial and topographical system which the artist must respect. For a painting to be deemed a 'proper law one', its elements must be in a correct proxemic arrangement. Scale and relationship are conventional but obviously in terms quite distinct from western linear writing.

The adage 'the map is not the territory' may not prove accurate for the Warlpiri. Where the land, a rockface or even a body is embellished, the embellishment sometimes may be an articulation of actual or perceived features of the thing itself. This is one area in which the transition to portable painted surfaces like canvas took some working out. Early acrylic paintings by traditional artists, for example, do not reduce large designs in scale to smaller surfaces. To achieve complexity in a painting, very large canvasses had to be provided. Small canvasses tended to depict what proved to be only small sections of large design complexes, like a single window in a computer spreadsheet. Only after some time and experience is an artist likely to scale down 'large' designs to smaller, marketable surfaces.

Because the 'icon', the inscription medium and the surface are all meaningful semiotic elements to be considered in the reading of Warlpiri graphics, it begins to appear that the system contains too many functioning semiotic levels to achieve either the denotation of pictographs or the abstract creativity of alphabets. But there are more levels than this. Eight are listed by Dubinskas and Traweek. (The additional ones are sociolinguistic matters of the performance event in which the design is produced and used.) We have now a problem like the one Bateson considers exemplified by the croquet game in *Alice in Wonderland*.[6] If the balls, the wickets and the mallets are all independently motivated moving elements, the game is too unpredictable to play. It is therefore an organic system, not a mechanical one, he concludes. Writing, if it is to accomplish the functions it achieves in the western world, must be a mechanical system, rule-bound and predictable. Reading must become automatic, transparent, 'the purely passive instrument of the spoken word'.[7] The system must detach itself from its human author and operate more or less equivalently for all literate users. This is precisely what Warlpiri graphics do not attempt. Warlpiri designs remain attached to their authors, as their property, and to paint another's design may be regarded as theft of a particularly onerous kind.

To call the owners of these designs their authors is not precisely correct, however. Warlpiri cosmology is staunchly conservative. It insists that truth pre-exists human apprehension. The creation and re-creation of the world is an established eternal process; the stories, songs, dances and designs which contain and explain these truths are likewise unchanging. What one paints or sings or dances is what your fathers and mothers and their fathers and mothers painted, danced and sang before you (although one may acquire additional information about these truths through revaluation or exchange). Your rights to do the same are determined by your position in an elaborately structured system of kin, itself handed down along with these expressive arts from the ancestors themselves.

In terms of what we know of cybernetic systems, this is impossible, a prescription for total cultural entropy. To explain how novelty enters the system, so that it can respond to environmental circumstances and remain viable, we have to step outside the explanations the system offers us.

Ethnology takes this vantage point when it notes ways in which the system does respond to change. Hale has described how pedagogic events may permit creativity. Wild has discovered ethnomusicological equivalents. Morphy has described the public display of secret designs. Rose has documented an historical account of Captain Cook becoming a legend, and I have analysed elsewhere elements of a history transforming to dreaming construction in which videotape has become involved.[8] What we have not so readily pursued is how our own activity in inscribing these in literate print may prove subversive of these traditions.

It is precisely the point that Warlpiri graphics are a writing system which does not subvert the authority of living people, or permit the identification of historical change as western literacy does. Warlpiri graphics oppose publication and public access. They do this partly by limiting denotation, operating instead as an evocative mnemonic, recalling stories without asserting any authorised text or privileged reading. They also do this by restricting access to both the production and viewing of designs. In the 'sand stories' described by Munn, which are women's casual accounting of dreams, daily life and

115

folktales, the public designs in this ephemeral, shifting medium are mostly illustrative and serve as entertainment and to introduce children to the system. In ritual, sacred and often secret designs are applied to bodies, objects and the ground, and the rules governing the construction and viewing of these designs are highly constrained, and permit less creativity. Most of these designs are obliterated following the ritual performance. Kolig observes that the secret designs painted on bodies may remain dimly visible in camp for days afterwards, undecipherable to the uninitiated, but testimony to the existence of secret lore nonetheless.[9] It would seem that the most sacred designs, those inscribed on sacred boards (*tjuringas*) and rock faces are also the most permanent. They are tended, and rock paintings and boards are ritually renewed cyclically. These are the only permanent texts, and their locations are rigorously guarded. For uninitiated or otherwise inappropriate people to view these, even accidentally, is punishable by death. Other than these, Warlpiri graphics are mentally stored, sociologically guarded and emerge only in ceremonial and story-telling performances.

These constraints on ownership and meaning become clear when, in the Yuendumu Artists Association studio, we try to identify paintings in preparation for sales. Usually, people refuse to provide the stories for any paintings but their own, or sometimes for designs owned by their kin group. However, understanding our relative naiveté and the economic incentive, sometimes when the painter is out of the community, another senior man or woman may try to provide a story for the painting. Remarkably, they sometimes fail, and more often come up with a reading which proves at variance with the 'correct' version provided later by the painter.

For a full, 'correct' version, a male painter will assemble other senior men of his patrimony associated with the same or adjacent land as himself, and negotiate the interpretation along with a set of men from the opposite patrimony who maintain rights as 'witnesses' or caretakers (*kurdungurlu*), but not performers or owners of the story.[10]

Glenn, in an insightful but overlooked note, provides an explanation.[11] Comparing Munn's early reports of Warlpiri iconography to 'State Department Graphics', he suggests that Warlpiri symbols seem much like the unique shorthand that translators and state department staff take of speeches during top-level meetings. These provide a means of recall of speech texts to be transcribed shortly after the event and prove quite accurate. But despite great similarities in the symbols used, rarely could a reporter reconstruct a speech from another's notes. Even the author might have difficulty as time elapsed. The difference between these ideolectic glyphs and Warlpiri symbols is partly in the collectivity of the Warlpiri system, where shared meanings are emergent in the interpretive negotiations that occur in graphic display events very much more than in the text itself.

Munn described these iconographs as devices for mediating symbolically between essential dualities expressed in Warlpiri cosmology, including subject/object, individual/collectivity, and especially dreaming time/present time. In the last instance, my analysis suggests there is more than a structuralist abstraction operating here: this writing system functionally resolves a conservative ideology, where things are always the same, with the contradictory lived evidence of a changing world. If this proves true, then the changes imposed on Aborigines since European invasion must present the most fundamental challenge to this system, and we begin to suspect that the dismal literacy rates represent more than just a failure of effort or will on the part of teachers or students. The failure of literacy might be seen as a resistance as well, not to change, but a threat to the culture's capacity to manage change.

Certainly, many senior people feel this way. They complain about the school and about writing. They say it makes the kids 'cheeky', so they don't listen to their elders. Especially interesting is the attitude of the middle-aged people who succeeded at the school in the 1950s–60s, before the bilingual programme was established. They say that the bilingual programme is holding up the kids, wasting their time with Warlpiri when they should be learning English, presumably as they did, usually by rote. (This contradicts all of the evaluation and testing evidence collected at Yuendumu which demonstrates that English is being learned better and more quickly in the bilingual programme.) Because this generation now operates in a uniquely powerful role in the community as spokesmen and leaders of the council and other European-inspired institutions (and thereby commands and channels substantial resources), they have been capable of engineering a considerable political challenge to bilingual education. The fact is, they themselves usually cannot read Warlpiri. These community leaders often are themselves in an anomalous cultural situation, and probably take the brunt of 'civilisation' – *ennui*, alcoholism, corruption – harder than anyone else. But from a distance, what we observe is a classical Aboriginal situation: generations in competition with each other. What is new is that the resource which the society afforded the most senior people to control – intellectual property in the form of dreaming law – is now subverted by the intellectual property accessible to younger people, through the medium of literacy.

TV and Video
In the last few years, another medium of inscribing and accessing information has entered Warlpiri life: video and television. Within a

year of the first VCR coming to Yuendumu (and sufficient tape rental services opening in Alice Springs to support these), videotape penetration was effectively total in the community. Only nine VCR's were reported in Aboriginal camps in 1983, for a population of 900, but these meant that essentially every extended family had access to at least one machine and could view in appropriate groupings, respecting avoidance restrictions and other traditional constraints on congregating. (When similar contents were screened as films at the school or church twenty-five years earlier, mass viewing required violations of traditional avoidance and association rules which produced considerable stress and often ended in chaos or fighting which hardly encouraged the acquisition of cinema literacy.)

My Aboriginal associate and I surveyed the situation and discovered, among other things, that it was costing at least $5,000 annually to maintain a video here, exceeding annual per capita income.[12] The difficulties of equipment purchase, maintenance, and programme supply in desert camps three hundred kilometres (along a mostly dirt road) from the nearest repair or tape rental shop proved enormous. Indeed, we found communities without electricity where ingenious generator installations were rigged up for the first time, not for lights or heat, but to play video. By the summer of 1985, the glow of the cathode ray tube had replaced the glow of the campfire in many, many remote Aboriginal settlements. There could be no question of motivation. Only rifles and 4-wheel drive Toyotas had achieved such acceptance of all of the introduced Western technologies.

The situation was reported with predictable alarm by the press, but also by Aborigines themselves (noting no contradiction in the fact that many of the most articulate objectors were themselves video watchers/owners). People were predicting culturecide and claimed that Aborigines appeared helpless to resist this new invasion. The analogy to alcohol was quickly made, and the received wisdom was that the electronic damage was already done, the pristine culture ruined, etc. These 'facts' were used with great ingenuity to support a wide and contradictory array of agendas, schemes and proposals, none, as far as I could tell, having much to do with the actual situation.

During this time I was watching Aborigines view video and watching them make video. It became quite clear that here was a situation where the bankruptcy of the 'effects' fallacy was demonstrated. From my observations, a very different 'uses and gratifications' picture emerged than the passive victimisation of Aboriginal audiences suggested by the alarmists.

One kind of evidence was available by asking my associates what any given videotape was 'about', in the most usual, conversational way. This produced what were to me, quite extraordinary readings of Hollywood programmes I thought I was already familiar with. Additional evidence came from school children and creative writing exercises. Another kind of evidence came from assisting in the production of indigenous Warlpiri video programmes along the lines of Worth and Adair's studies of Navajo film-making and analysing both production style and product.[13] The evidence from these two sources proved complementary, and permitted the beginnings of a theory of Aboriginal interpretation of imported television. Here, I will summarise some of the elements of this theory, which I have treated more fully elsewhere.[14]

The most suggestive finding was that Aboriginal people were unfamiliar with the conventions, genres and epistemology of western narrative fiction. They were unable to evaluate the truth value of Hollywood cinema, to distinguish, for example, documentary from romance. This may be because all Warlpiri stories are true, and the inscription and interpretation processes which assure their preservations also ensure their truthfulness (or at least engineer a consensus on what is true at each re-creation, which amounts to the same thing). Thus, I was observing the impact of fiction on Aborigines much more than the impact of television *per se*.

Comparisons between Warlpiri story form and imported video fictions demonstrated that in many instances content (what is supplied in the narrative) and context (what must be assumed) are so different from one system to the other that they might be said to be reversed. For example, Warlpiri narrative will provide detailed kinship relationships between all characters as well as establishing a kinship domain for each. When Hollywood videos fail to say where Rocky's grandmother is, or who's taking care of his sister-in-law, Warlpiri viewers discuss the matter and need to fill in the missing content. By contrast, personal motivation is unusual in Aboriginal story; characters do things because the class (kin, animal, plant) of which they are a member is known to behave this way. This produces interesting indigenous theories, for example, of national character to explain behaviour in *Midnight Express*, or *The A-Team*. But equally interesting, it tends to ignore narrative exposition and character development, focusing instead on dramatic action (as do Aboriginal stories themselves).

Violence, for instance, described by Gerbner as TV's cheap industrial ingredient overlaid on narratives mainly to bring viewer's attention to the screen for ultimately commercial purposes, is for Aboriginal viewers the core of the story. The motivation and character exposition which the European viewer is expected to know to explain why someone was mugged or robbed or blown up is missing. It is more likely in Aboriginal accounts to be supplied by what we

119

would consider supernatural reasons, consistent with the reasons misfortunes befall people in Aboriginal cosmology.

These brief examples should make it clear that it will be very difficult to predict the effects of particular television contents on traditional Aboriginal audiences without a well-developed theory of interpretation. To advance a theory of interpretation, it may be helpful to consider now what kind of writing system is television which makes it so accessible and attractive to these people whose traditional preference was for writing systems of the sort explained above? An analysis of the videotapes the Warlpiri made themselves provides some insight.

Warlpiri videotape is at first disappointing to the European observer. It seems unbearably slow, involving long landscape pans and still takes that seem semantically empty. Much of what appears on screen, for better or worse, might easily be attributed to naive film-making. Finally, the entire corpus of 300 hours of tape shot at Yuendumu is all documentary-type 'direct cinema'. The only exceptions, when events were constructed and people performed expressly for the camera, were the result of my intervention for experimental purposes.

Yet Warlpiri audiences view these tapes with great attention and emotion, often repeatedly, beyond what could be expected from a fascination with 'home movies'. If this is attributable to novelty, it has yet to wear off in the three years I've been involved. In fact, the limits to some tape's lifespan are the limits of the lifespan of the characters. A mortuary rule prohibiting the mention of the names of dead people now applies as well to their recorded images, which means that when a tape contains pictures of the deceased, that tape is no longer shown.

Producers and viewers will describe the tape, its purposes and meanings in ways not immediately apparent from the recorded images themselves. For example, proper videotape production for a particular story may require the presence of several families including many people. But not only do most of these people not appear on the tape, but a proportion of them (related to the on-screen 'owner' of the story through the mother's lineage) must not appear on screen. They may, however, operate the camera. This is consistent with equivalent rules in ceremonial performance. But what attracts our attention is that everybody seems to know how that tape was made, and whether these rules were observed, and therefore if the tape is a 'proper' and 'true' story, without any apparent evidence on the tape itself. Similarly, what are to the European observer semantically empty landscape pans are explained by Aboriginal producers and viewers as full of meaning. The camera in fact traces tracks and locations where ancestors, spirits or historical characters travelled.

The apparently empty shot is quite full of life and history to the Aboriginal eye. The electronic inscription process may be said to be operating for the Warlpiri in a way not unlike their graphic system, providing mnemonic, evocative symbols amenable to interpretation and historical accuracy when viewed in the proper social and cultural context.

Video viewing is a very active interpretive social event, particularly in groups, made possible by private ownership of VCR's. In fact, ownership of VCR's is not quite private in the European sense, and our early survey determined that these machines were corporate purchases and circulated along traditional exchange/obligation routes, usually within families. This means that viewing groups, as described above, are appropriate assemblages of kin, and provide a suitable setting for events of social interpretation, not unlike what was described for Warlpiri artists explaining their paintings. One interesting difference between groups viewing Warlpiri productions and those viewing European ones is the placement of the elders. With the former, elders sit towards the front, turned half around to interpret to the younger people. With Hollywood videos, the children sit in front, often interpreting and explaining to their elders.

Conclusions

Warlpiri Aborigines have perhaps discovered some comfortable analogies between their experience of video production and viewing and their own traditional graphic system. For Warlpiri viewers, Hollywood videos do not prove to be complete, authoritative texts. Rather, they are very partial accounts which require a good deal of interpretive activity on the part of viewers to supply contents as well as contexts which make these narratives meaningful. When home video made it possible for Warlpiri to control the place and membership of viewing groups, it became possible to assemble the small, interpretive communities which are associated with other performances in which stories are told and their associated graphics displayed. At this point, video viewing became a most popular and persuasive camp activity.

By contrast, reading and literature did not. While this article has not expanded a theory of literacy except by contrast to Warlpiri graphics, an historical point is worth making. Current literary criticism now questions the notion that there are correct or even privileged readings of literary texts; reader-centred criticism claims that readers will make diverse interpretations. But the Warlpiri were introduced to literature through that most privileged of texts, the Christian Bible, by missionaries who took the Calvinist position that every word therein had one and only one 'common sense' meaning. Thus, literature may well be associated by Aborigines with a dogmatism, a certitude, a sense of revealed and inscribed truth which would prove subversive of their own dreaming history and law. That history and law required a different mode of writing to maintain the continuity of its authority.

The evidence from Warlpiri graphics and Warlpiri video prove useful to the re-examination of the 'oral/electronic' analogy proposed by my media historians. It would be important in this re-examination to note first that some 'oral' societies, for example the Warlpiri, do have writing of a particular sort: a writing subservient to, and in the service of, oral performance and living authorities. The writings which became the source of western literacy are distinguished functionally from 'oral writing' in that they subverted and replaced orality as the integrating mode of symbolisation for society.

We may now question whether written texts are in fact so authoritative as we have been used to considering them. But we raise this question at a time when another inscription system competes for centrality in our information processing and imaginative symbolising: electronics. It seems particularly interesting that this new inscription process is proving more accessible, and perhaps less culturally subversive, to people in those remaining enclaves of oral tradition. The most useful first question to ask may be, 'what about their writing systems is like electronic writing?' rather than 'how is their society like ours?'

At least one result of this kind of inquiry is to suggest some useful considerations for dealing with commercial video narratives and emergent television genres. Although historically the sources for plot and character are found in earlier novel and literary forms, Warlpiri viewers, and perhaps now many others, put video fictions to quite different uses and make quite different sense of them. What Warlpiri viewers require is a good deal of visual and visceral action, a rich familial and kinship context and a means of combining these into a classificatory universe whose truth is partly in the structures they can produce with these elements and partly in the opportunities the texts provide for negotiation and social discourse. As western television develops its own conventions, themes and genres, reaching out for the vastest, pan-cultural (or a-cultural) mass audiences, it is clearly offering more of these kinds of materials to its viewers. To the horror of literate-biased critics, it is stripping away cultural denotation, cultural specific motivations, psychologies and their place in character development. Its commitment to 'inscribed truth' is gone. Indeed, we might seem to be realising Lévi-Strauss's prophecy:

> ... this universe made up of meanings no longer appears to us as a retrospective witness of a time when: '... *le ciel sur la terre marchait et respirait dans un peuple de dieux,*' and which the poet

evokes only for the purpose to ask whether it is to be regretted. This time is now restored to us, thanks to the discovery of a universe of information where the laws of savage thought reign once more: 'heaven' too, 'walking on earth' among a population of transmitters and receivers whose messages, while in transmission, constitute objects of the physical world and can be grasped from without and from within.[15]

That video now proves acceptable and accessible in a way that alphabetic literature did not, could prove to be partly a transitional feature of Aborigine's recent encounter with the medium. Warlpiri-produced video may preserve these features, if its unique properties are recognised and encouraged (a difficult proposition when over-zealous 'professional' trainers intercede, and media institutions force competition between indigenous and imported programmes). But it is likely that Warlpiri people will develop greater sophistication in European genres and the interpretation of imported narrative fiction which will bring them closer to European readings. In so doing, it may be that the 'gaps' I have identified will fill up, and the medium will become more denotative and less evocative, and finally, less Warlpiri.

It could prove promising that the most popular genres appear to be action/adventure, soaps, musicals and slapstick. Whatever our educated palates may think of these forms, they have advantages in the context of this analysis. As the least character motivated, most formulaic fictions, they may encourage active interpretation and cross-culturally varied readings. The trend in popular TV and international video marketing continues to be in favour of those entertainments in which universal familial relationships are highlighted, action is dominant and culture specific references are either minimal or unnecessary for the viewer's enjoyment. From this perspective, it would seem difficult to see in the introduction of imported video and television programmes the destruction of Aboriginal culture. Such a claim can only be made in ignorance of the strong traditions and preferences in graphics, the selectivity of media and contents and the strength of Warlpiri interpretation. Such ignorance arises best in unilineal evolutionism.

References

1. E. Michaels, 'Constraints on Knowledge in an Economy of Oral Information', in *Current Anthropology*, vol. 24, October 1985.
2. N. Munn, 'The Transformation of Subjects into Objects in Warlpiri and Pitjantjatjara Myth', in R. Berndt (ed.), *Australian Aboriginal Anthropology* (Nedlands: Univ. Western Australia Press, 1970), and N. Munn, *Warlpiri Iconography* (Ithaca: Cornell Univ. Press, 1973).

3. F. Dubinskas, S. Traweek, 'Closer to the Ground: A Reinterpretation of Warl-piri Iconography', in *Man*, vol. 19 no. 1, 1984.
4. E. Havelock, *The Literate Revolution in Greece and Its Cultural Consequences* (Princeton, N.J.: Princeton Univ. Press, 1982).
5. C. Lévi-Strauss, *The Savage Mind* (Chicago: Chicago Univ. Press, 1966).
6. G. Bateson, 'Why Do Things Have Outlines?', in Bateson, *Steps to an Ecology of Mind* (New York: Ballantine, 1972).
7. Havelock, *Literate Revolution*, p. 55.
8. K. Hale, 'Remarks on Creativity in Aboriginal Verse', in J. Kassler, J. Stu-bington (eds.), *Problems and Solutions* (Sydney: Hale and Iremonger, 1984); S. Wild, *Warlpiri Music and Dance in their Social and Cultural Nexus*, unpub-lished doctoral dissertation, Univ. Indiana, 1975; H. Morphy, 'Now You Understand', in N. Peterson, M. Langton (eds.), *Aborigines, Land and Land Rights* (Canberra: Australian Institute of Aboriginal Studies, 1983); D. Rose, 'The Saga of Captain Cook', *Australian Aboriginal Studies*, vol. 2, 1984; E. Michaels, F. Kelly, 'The Social Organisation of an Aboriginal Video Work-place', *Australian Aboriginal Studies*, vol. 1, 1984.
9. E. Kolig, *The Silent Revolution* (Philadelphia: Ishi, 1981), p. viii.
10. D. Nash, 'An Etymological Note on Warlpiri Kurdungurlu', in J. Heath et al., *Languages of Kinship in Aboriginal Australia* (Sydney: Univ. Sydney, 1982).
11. E. Glenn, 'Warlpiri and State Department Graphics', *American Anthropolo-gist*, vol. 65, 1963.
12. E. Michaels, L. Granites, 'The Cost of Video at Yuendumu', in *Media Informa-tion Australia*, vol. 32, 1984.
13. Michaels and Kelly, 'Social Organisation'. See S. Worth, J. Adair, *Through Navajo Eyes* (Bloomington: Indiana Univ. Press, 1972).
14. E. Michaels, 'The Indigenous Languages of Video and Television in Central Australia', paper presented to the Conference on Visual Communication, Annenberg School, Philadelphia, 1985.
15. Lévi-Strauss, *The Savage Mind*, p. 267, quoting de Musset.

PAULA MATABANE

Subcultural Experience and Television Viewing

Television has been described as a 'cultural forum' reflecting various themes in society which are subjected to selective readings or interpretations by viewers.[1] Rejecting the informative sender-message-receiver model of communication, many researchers now argue that culture is an intervening variable through which media content is filtered and interpreted. The failure to identify the activity and sociocultural conditions of the receiver has caused a crisis in the dominant audience paradigm, resulting in the appearance of 'anomalous' sections of the audience. This is particularly true of the way mainstream research has characterised the viewing behaviour of the black audience.[2] The overall purpose of this study is to investigate the relationship between viewing and subculture. Specifically, the study will examine the relationship between different types of viewing as evidence of viewer selectivity, and the structural relationships between television viewing and experience derived from subculture and location in the social structure.

Before describing the data of the study it is necessary to review the literature on viewing and on subcultures. Much has been written about the nonselectivity of television viewing especially among heavy viewers. Noll argues for the passive viewer model, where viewers watch the medium of television rather than specific programmes, and where the true choice is between viewing and not viewing.[3] Gerbner also assumes that audiences are relatively nonselective, but contends that this behaviour is encouraged by the homogeneous nature of content across all genres with respect to violence, social relationships, value and outcomes.[4]

These positions are contradicted by the first-choice-only model, in which the potential audience is divided into subgroups with internally homogeneous programme preferences.[5] Philport further suggests the need to distinguish viewers' predisposition to select a programme based on its format or own unique characteristics.[6] However, research by Allen strongly supports the position that television viewing by the black audience is a selective process dependent upon racial experience and race-related psychosocial factors or predispositions, despite the disproportionate percentage of black viewers who are

125

heavy television viewers and hence presumed to be even more nonselective.[7]

White has argued that researchers should focus on the situational conditions that determine when a receiver will 'select' information from a source – 'selection' implying that the receiver perceives the information as relevant.[8] If this view is correct, then the viewing of television content is problematic for black audiences, given the presence of racial and racist stereotypes and the low representation of black images.[9]

If there is selectivity, what is the basis for it? We must examine the very concept of television viewing, which can be understood on at least three levels: viewing irrespective of content, viewing specific content, and involvement with specific content.[10] Furthermore, as Hawkins observes, 'to the extent that we perceive television's content to be a realistic portrayal of life, we may be more affected by and learn more from that content'.[11] Comstock and colleagues also conclude that perceived reality is a necessary condition for any relationship of influence between viewing and behaviour to occur, irrespective of degree of exposure or consequential values associated with the programme.[12] It follows that to the extent black viewers evaluate content differently with respect to its true-to-lifeness, one may observe viewer selectivity, and it is reasonable to expect social experience to be an important factor in determining viewer perceptions of content realism.

Earlier we noted the intensive stereotyping that is characteristic of black character portrayal on television. Content analysis suggests there are differences among programmes which may offer viewers a basis of selection. Black characters tend to be cast either in an all-black setting or as the lone black person in an otherwise all-white setting; the all-black settings are usually portrayed in a low-income environment with few socially productive persons and/or themes; the stereotyping of black characters in an all-black setting is intensified through the use of black English, and black English is more often used by low income comic characters.[13] Black characters are cast predominantly in situation comedies with storylines that avoid discussions about race or racism and poverty.

In this study subculture is understood as a particularised configuration of historically evolved selected processes through which individuals channel their reactions to internal and external stimuli. It provides the individual with the social and cultural frameworks through which they experience life and obtain their important symbolic and real world environment. Key subcultural experiences for Afro-Americans are derived from race of neighbourhood, tenure of neighbourhood residency, age, church-going, frequency of church attendance, race of church denomination attended, and community participation. Important social structural factors include sex, income, education, home ownership, occupational status, and subjective social class.[14]

Methodology
Data for this study was collected by a telephone survey of 161 adult Afro-Americans between 1 October–7 November 1984 in Washington DC. Interviews were conducted in four census tracts selected on the basis of income level and racial composition. This represents a limitation on the generalisability of the study to an initial investigation of relationships.

The sample was selected from two predominantly black neighbourhoods (over 90% black), one racially mixed neighbourhood (50% black), and one predominantly white neighbourhood (20% black) The first available adult per household was interviewed. Telephone numbers were obtained from a reverse telephone directory and randomly assigned a calling order. A relatively low response rate of 22% was obtained due to a large number of not-at-home responses and lines changed. Follow-up calls for lines changed were not made. Pearson's partial correlations were used to test for a relationship between the three different viewing variables.

Factor analysis using varimax rotation was used to explore the possibility of interrelations among programmes rated by viewers for perceived reality. Four programme factors were observed based on factor loadings of individual programmes. Averaged composite index scores for each factor were constructed using the raw scores of respondents on each programme. In addition, one overall composite index score was similarly constructed based on raw scores on all the programme items rated for perceived reality. We were interested in observing differences in structural linkages between different types of experiences and programme factors based on the step-wise hierarchical regression analyses.

In the first set of regression analyses, general television viewing, black character programme viewing (composite), daily black character viewing, weekly black character viewing and the overall composite of perceived reality of black character programmes (alpha = .81) were each regressed against the experience variables in a step-wise manner. In the second set of analyses, the four perceived reality progamme factors were each regressed, first on weekly black character viewing, daily black character viewing and general viewing, then regressed on the experience variables. The four factors consisted of:

a) all-black setting (*The Jeffersons*, *What's Happ'ning*, *Good Times* and *Sanford and Son*) (alpha = .77)

127

b) integrated setting (*Webster, Gimme a Break, Dif'rent Strokes,* and *Benson*) (alpha = .73)
c) urban realism setting (*Hill Street Blues*)
d) macho fantasy setting (*The A-Team*)

The model for this study included television viewing variables and variables measuring subcultural experience and location in the social structure. Television viewing was measured using three separate self-report variables. Firstly, the viewer's estimate of how many hours he/she spent watching television yesterday; secondly, the viewer's estimate of the number of episodes of eight weekly black character programmes she/he watched in the past four weeks and the number of episodes of four daily black character programmes he/she viewed in the past week. This score was summed with one point given for each episode of any programme viewed. The third variable was the viewer's perceived reality of the black character programmes listed above. Respondents were asked to evaluate these programmes for their perceived 'true-to-lifeness' using a ten-point interval measure scale. Higher scores indicated a higher level of perceived reality. Scores were averaged.

Respondents' participation in the black subculture was measured by race of neighbourhood (coded from census data), tenure of neighbourhood residency, church-going, frequency of church attendance, race of church denomination, community participation and age.[15]

Respondents' location in the social structure was measured by sex, education, income, subjective social class, home ownership, and occupational status.

Findings
The correlational analysis suggests the respondents in this study were engaged in different types of television viewing in a manner consistent with our theoretical notions concerning viewer selectivity. Table I shows a moderate relationship between general viewing and

TABLE I

Relations between television viewing variables (Pearson's)

	BCV	BCE
Black Character Programme Viewing (BCV)		
Perceived Reality of Black Character Programmes (BCE)	.34***	
General TV Viewing (GTV)	.29***	.07

*p<.05   **p<.01   ***p<.001

128

black character programme viewing (r = .29, p < .001), a stronger relationship between black character programme viewing and perceived reality of black character programme (r = .34, p < .001). There is no relationship between perceived reality and general viewing. Partial control for perceived reality has no effect on the bivariate relationship between general viewing and black character viewing. Similarly, partial control for general viewing has no effect on the bivariate relationship between perceived reality and black character viewing.

The factor analysis with varimax rotation revealed four perceived reality programme factors designated as follows: all-black setting, integrated setting, urban realism setting, and macho fantasy setting (see Table II). None of the programme variables loaded heavily on more than one factor, thereby suggesting orthogonality.

TABLE II

Perceived reality of black character programmes factor loadings

	Loadings
Factor 1: All Black Setting	
The Jeffersons	.58
Sanford and Son	.57
What's Happ'ning	.59
Good Times	.76
Factor 2: Integrated Setting	
Gimme a Break	.50
Dif'rent Strokes	.59
Benson	.70
Webster	.54
Factor 3: Urban Realism	
Hill Street Blues	.62
Factor 4: Macho Fantasy	
The A-Team	.92

In the first set of regression analyses, structural linkages between the experience variables and general viewing, black character viewing and perceived reality were tested. The experience variables explained 15% (p < .01) of the variance in general viewing, 36% (p < .01) of black character viewing and 26% (p < .01) of perceived reality (see Table III). Black character viewing was disaggregated into daily and weekly viewing. The experience variables explained 34%

TABLE III
Regression models predicting television viewing

| Conditional variables | GTV | Criterion variables | |
		BCV	BCE
Education	−.27*	−.29**	−.36**
Community participation	−.15	−.13	.16
Home ownership		−.29**	−.21*
Subj. social class			.17
Black church	−.23*		
Church-going	.17		
Freq. church		.21*	
Generation		.25*	
Income		−.11	−.17
Occupational status			
Sex	.11		
Black neighbourhood			
Tenure residency			.15
R-squared	15%	36%	26%
s.e.	1.9	9.4	1.6
F	3.4**	9.0**	5.6**

*p<.05   **p<.01

(p < .01) of daily black character viewing and 33% (p < .01) of weekly black character viewing (see Table IV).

A path analytic model of the experience variables provides the most parsimonious explanation of general viewing (Table III). Education and black church attendance make the largest contributions to explained variance and are the only two significant variables. Both are inversely related to general viewing as is community participation. Notably, church-going makes a positive contribution to general viewing although black church attendance is negative. For this sample, a heavy general television viewer is a black female of low education, and with low community participation except for her attendance at a non-black church. She is also a long-term resident of her neighbourhood.

In explaining black character programme viewing, while education, community participation and tenure of neighbourhood residency overlap from the general viewing path model, the sign for community participation is inversed and education makes a larger contribution to explained variance. Further, we observe three new experience variables in this model that were not in the general viewing model – subjective social class, home ownership and income. The heavy viewer of black character programmes is a long-term resident of his neighbourhood and has a low education and low income. He or

130

she is not a home-owner but still identifies as middle-class or above. This viewer is also active in the community.

Education and home ownership make the largest contributions to explained variance of the perceived reality of black character programmes and both are inversely related to viewing. Generation and frequency of church attendance are significant explainers of perceived reality, though not for the other two types of viewing. Community participation and income are inversely related to the perceived reality.

When black character programme viewing is disaggregated into daily and weekly scheduling patterns there is a differentiation among the variables explaining viewing (see Table IV). While edu-

TABLE IV

Regression models predicting daily and weekly black character programme viewing

| | Criterion variables | |
| | Black character programme viewing | |
Conditional variables	weekly	daily
Education	−.24*	−.53**
Freq. church	.27**	
Age	.30**	
Home ownership	−.24*	−.26*
Income	−.15	
Residence tenure		.18
R-squared	33%	34%
s.e.	5.97	4.86
F	9.47**	15.7**

*p<.05   **p<.01

cation is inversely linked in both cases it is far more important in predicting viewing of daily programmes. Home ownership, too, is related to both but there are no other variables that predict viewing of weekly and daily programmes.

Daily black character viewing is explained by three experience variables: education, home ownership and tenure of neighbourhood residency (which, however, is not significant in the equation). The viewing of weekly black character programming is a more complex variable. In addition to education and home ownership, it was also explained by age, frequency of church attendance and income (which was not a significant predictor). For both daily and weekly black character programming, subcultural variables increased the likelihood of viewing, while increases in location in the social structure predicted decreases in viewing.

TABLE V

Regression models predicting perceived reality of black character programme factors

Conditional Variables	All Black	Integrated	Urban Realism	Macho Fantasy
		Criterion variables		
Daily BC viewing	.23*			
Weekly BC viewing		.31**		.36**
General viewing		−.17	.19	
Age	.28**		−.35**	
Occupation	−.23*	−.21*		
Black church	.18			
Education		−.25*		−.23
Community part.		.13		.16
Sex (female)			−.27**	
Income			.17	
Black neighbourhood			.15	.14
Residence tenure				−.24*
R-squared	22%	23%	19%	25%
s.e.	2.02	1.91	3.0	2.29
F	7.12**	6.08**	5.0**	6.66**

*p<.05   **p<.01

Table V indicates that the experience variables differentiated according to different programme factors with regard to the perceived reality of black character programmes. The experience variables and television exposure variables (general viewing, weekly black character programme viewing, and daily black character programme viewing) explained 22% of variance in all-black settings, 23% in integrated setting, 19% in urban realism and 25% in macho fantasy settings.

A high perceived reality of the all-black setting programmes tends to be associated with an older person who views daily black character programmes heavily. These individuals are of low occupational status and attend a black church. A high perceived reality of the integrated setting programmes was most characteristic of persons with a low education and low occupational status but active in the community. These individuals view weekly black character programmes heavily but are not heavy general viewers of television, suggesting selectivity.

Urban realism was seen by young males with a high income as more realistic. They tend to live in black neighbourhoods and view general television heavily. A high perceived reality of macho fantasy programmes was characteristic of heavy viewers of daily black character programmes. These viewers had a low education, were active in the community and were short-term residents of a black neighbourhood.

Discussion and Conclusion

An often heard criticism of television studies is the inadequate operationalisation of the viewing measures. In Table I, the intercorrelations of our three viewing measures suggests each may be tapping a different construct of 'viewing'. The moderate size of the relationship between general viewing and black character viewing (r = .29) implies that only part of general viewing consists of black character viewing. One form of viewing does not strongly imply the other. Further, while black character viewing and perceived reality of black character programmes are also moderately related (r = .34), perceived reality of black character programmes and general viewing are not related. Hence, belief in one type of content does not translate into an overall acceptance of television content as realistic.

For our sample, these findings may suggest that black viewers distinguish between black character and nonblack character programmes. They may perceive at least two worlds on television – one black, the other white. Perceptions or evaluations of one world do not necessarily carry over into the other. In short, our viewers may be selective. An examination of the experiences that best predict or explain variance in each type of viewing supports this notion of selectivity. Education and community participation were common predictors for general viewing, black character viewing and perceived reality of black character programmes. However, each viewing model also included experience predictors exclusive to that model alone.

General television viewing was best explained by education, black church attendance, church-going, community participation and sex. A heavy general viewer was likely to be female and attend church. In most studies activities outside the home are seen to decrease television viewing, but in this study, we observe that two measures of the same type of activity have opposite influences on viewing. That is, black church attendance also decreases the likelihood of viewing general television. Clearly, the content of other activity influences the content of television which is sought. Perhaps there is no one overall relationship between experience and viewing, but the relationship is specific to type of content viewed. This belief is supported by changes in the type of contribution (negative or positive) that community participation makes to different types of viewing. Community participation is inversely related to general viewing. That is to say, the overall or total amount of time spent viewing is decreased as one is more active in the community. But this relationship does not hold up for all types of viewing as we will observe below.

Black character viewing (composite) was best explained by education, home ownership, generation, frequency of church attendance, community participation and income. Only generation and frequency of church attendance make a positive contribution to black character viewing. Again we observe that two activities which occur outside the home, in this case frequency of church attendance and community participation, have opposite correlational signs with viewing. Both activities consume time, but imply different behaviour toward race-specific content. We observe another difference in the type of experiences predicting general viewing as against black character viewing. For general viewing the only social structural variable explaining variance was education (gender is not a social structure variable *per se*), subcultural factors being the main variables. On the other hand, for black character viewing, social structure becomes more significant with three social structural variables explaining variance. This differentiation of types of experiences in each viewing model suggests that type of subcultural experience is the most significant factor (aside from education) to determine the viewing of black or nonblack character content. Within the black character category, location in the social structure becomes more significant in differentiating viewers, although subcultural experience is still significant.

The experiences predicting the viewing of daily and weekly black character programmes vary. Subcultural experience is more important in explaining the viewing of weekly than of daily programmes. Participation in the black subculture increases viewing but higher social structural location decreases viewing.

The perceived reality of black character programmes was best explained by education, subjective social class, income, home ownership, community participation and tenure of neighbourhood residency. Social structural location clearly dominates over subcultural experience in differentiating viewers. Education is the only significant predictor suggesting that the ability to decode television's messages is the most important factor. However, the presence of so many social structural variables strongly suggests that internal factors of these black character programmes are very important to viewers and that all viewing is not equal. Persons who identify as middle-class or above and are low-income non-home owners but active in their communty are most likely to believe in the realism of black character programmes. What is of much interest is that community participation is positively related to a viewing variable. This endorses earlier speculation about selectivity in viewing and the content of social experience. Community participation may reduce time for exposure to television but it increases one's belief in the realism of black character programmes.

The perceived reality variable is probably our most complex and important viewing variable in illustrating how television viewing interacts with experience. We observed earlier that black characters on television tend to be dichotomised into racial settings characterised by socio-economic status. The factor analysis of perceived reality confirms our belief that black viewers select black character programmes along these same dimensions.

Four perceived reality programme factors were observed in the analysis. The two dominant factors represent a clear racial dichotomisation between the all-black setting and the integrated setting. The path analysis suggests that type of subcultural experience differentiates the viewers of each factor who tend to be of low social status. Older people tend to find the all-black setting more realistic, but quite significantly, so do the black church-goers, while those with a high level of community activity find the integrated settings more realistic. Church-going and community were used to measure subcultural activity, and originally we expected much of religious activity to be included under community participation, but found no significant bivariate relationship between the two measures. These are clearly two different types of activity that predict preference for two varying types of television content.

A correlational analysis of the relationships among the experience variables used in this study suggests that persons with a high level of community activity tend to be of higher social status and live in a black neighbourhood. They are social status minorities within their own communities. The socio-economic indicators overall were inversely correlated with black neighbourhood residence. On the other hand, black church-goers tend to be older persons of low social status. The quality of racial experience along with socio-economic status appears to influence perceived reality of television content along racial dimensions.

The urban realism factor (as in *Hill Street Blues*) was reported as more realistic by young males living in a black neighbourhood and with higher income levels. They prefer general television viewing. This was the only perceived reality factor where a black character viewing variable was not a significant predictor of viewing. For this factor, though income differences are important, education is not. We speculate that these males may be skilled workers with a high school education, which explains their higher income levels, and may also explain their preference for an urban occupation-based drama over home-based situation comedies in the all-black and integrated setting factors. Further, this factor also featured an integrated setting, suggesting that high-income blacks reject the all-black setting as realistic.

The macho fantasy factor (*The A-Team*) was perceived as more realistic by black neighbourhood residents with a low education but who are active in the community. This factor also featured an integrated setting but the lead character, Mr T, has been severely criticised as perpetuating a stereotypical image in terms of his near ludicrous strongman costume and Mohawk Indian haircut. Thus the low-status community activist sees this integrated setting as realistic but his/her low education limits his ability to decode the message of racism.

These findings are preliminary and come from an exploratory study but they suggest new and highly significant insights into television viewing and social experience. The model presented here must be tested at a more representative level and include a multi-racial sample. The general viewing variable should be expanded to identify specific programmes viewed and their perceived realism. Community participation should be specified by type of activity in order to understand the types of activities and ideology underlying this variable. Notwithstanding the general limitations on this study and the need for caution in making conclusions, we feel that it indicates that television viewing is a selective process influenced by social experience derived from subculture and location in the social structure.

References

1. See H. Newcomb, P. Hirsch, 'Television as a Cultural Forum: Implications for Research', in W. D. Rowland, B. Watkins (eds.), *Interpreting Television: Current Research Perspectives* (Beverly Hills: Sage, 1985).
2. See *inter alia* G. Comstock et al., *TV and Human Behavior* (New York: Columbia University Press, 1978); R. A. White, 'Mass Communication and Culture: Transition to a New Paradigm', in *Journal of Communication*, vol. 33, 1983.
3. R. Noll et al., *Economic Aspects of Television Regulation* (Washington, DC: Brookings Institute, 1973).
4. See G. Gerbner et al., 'The "Mainstreaming" of America: Violence Profile no. 11', in *Journal of Communication*, vol. 30, 1980.
5. P. Steiner, 'Program Patterns and Preferences, and the Workability of Competition in Radio Broadcasting', in *Quarterly Journal of Economics*, vol. 66, 1952.
6. J. Philport, 'The Psychology of Viewer Program Evaluation', presented at CPB Technical Conference on Qualitative Ratings, March 1980.
7. R. L. Allen, 'The State of Communication Research on Black Americans', in H. A. Myriack (ed.), *In Search of Diversity: Symposium on Minority Audiences and Programming Research: Approaches and Applications* (Washington, DC: C.P.B., 1981); R. L. Allen and W. T. Bielby, 'Blacks' Attitudes and Behaviour Toward Television', in *Communication Research* 6, 1979; R. L. Allen and S. Hatchett, 'The Media and Social Effects: Self and System Orientations of Blacks', in *Communication Research* vol. 13 no. 1, 1986; A. C. Nielsen Company, Black American Study: October 1981–February 1982 (New York: A. C. Nielsen Company, 1982); C. Stroman and L. B. Becker, 'Racial Differences in Gratification', in *Journalism Quarterly*, vol. 55, 1978.
8. White, 'Mass Communication and Culture'.

9. See, for instance, P. Baptista-Fernandez and B. S. Greenberg, 'The Context, Characteristics and Communication Behaviors of TV Blacks', in B. S. Greenberg (ed.), *Life on Television: Content Analyses of US TV Drama* (Norwood, NJ: Ablex, 1980); F. J. McDonald, *Blacks and White Television* (Chicago: Nelson-Hall, 1983); United States Commission on Civil Rights, *Window Dressing on the Set: Women and Minorities in Television* (Washington, DC: Government Printing Office, 1977).

10. See G. Salomon and A. Cohen, 'On the Meaning and Validity of Television Viewing', in *Human Communication Research*, vol. 4, 1978.

11. R. Hawkins, 'The Dimensional Structure of Children's Perceptions of Television's Reality', ERIC ED 120 855.

12. Comstock et al., *TV and Human Behavior.*

13. See P. Baptista-Fernandez and B. S. Greenberg, 'The Context, Characteristics [...]'; C. Banks, 'A Content Analysis of the Treatment of Black Americans on Television', in *Social Education*, 41, 1977, pp. 336–39; and M. Fine et al., 'Black English on Black Situation Comedies', in *Journal of Communication*, vol. 29, 1980.

14. See K. Deutsch, *Nationalism and Social Consciousness* (Cambridge, Mass.: M.I.T. Press, 1966); D. Morley, *The 'Nationwide' Audience* (London: British Film Institute, 1980); J. N. McLeod and G. O'Keefe, 'The Socialisation Perspective and Communication Behavior', in F. G. Kline and P. Tichenor (eds.), *Current Perspectives in Mass Communication Research* (Beverly Hills: Sage, 1972).

15. Community participation was measured using a revised version of Miller's Scorecard for Community Services Activity (alpha = .75) [see D. Miller (ed.) *Handbook of Research Design and Social Measurement*, New York: David McKay, 1977]. Churches coded as black denomination were: Baptist, A.M.E., Holinecs, Pentecostal, Apostolic, Church of God, Church of Christ, and Evangelical Christian. Generation was coded as old (born before 1927), middle-aged (born 1928–1946) and young (born 1947–1966).

IV THE CHILD AUDIENCE

PATRICIA PALMER

The Social Nature of
Children's Television Viewing

Introduction

A powerful being which has children in its maw is the stuff of legends, evoking our desires to protect children from harm. If television is such a creature, we must roundly condemn it or at the very least exercise strict control over its access to children. Writing about children's television, for example, is full of calls for greater control – by parents, governments, anybody but children themselves. Research usually supports and encourages these calls for control, through its uncritical view of the child-television relationship as one of dream-like passivity.[1] Those studies which have directly challenged current definitions of the child-TV relationship do not seem to have made much of an impression on the prevailing view of the medium as nasty.[2]

The project described here took a different tack from the beginning. First, it used a conceptual framework with which television viewing could be studied as a process of communication rather than a one-way transfer of information or 'effects' from the set to the person. It was also decided that children's own definitions of the experience of television viewing should shape the direction and concerns of the research project.

This approach was based on the views of George Herbert Mead and others.[3] It was further developed by Blumer, whose writing included a critique of mass media research.[4] In this critique, Blumer exposed the assumptions of 'effects' research and set a new agenda for the study of mass communication. It was Blumer who coined the name 'symbolic interactionism', the perspective on which the following research project was based. Researchers following him have studied communication processes within and between social groups, including television audiences.[5] Recent formulations of this perspective have emphasised the significance of social context and social structure in ongoing group interaction,[6] factors which have previously been ignored or underplayed because of Blumer's insistence on the 'intertwined, interacting, and transforming make-up of the communicative process'.[7]

Symbolic interactionism is thus one variant of a more general constructivist approach.[8] The specific framework of symbolic interactionism was preferred because of its existing research literature on children and television.[9] In addition, researchers using this perspective have developed new methods with which to study social groups.[10] These methods usually combine observation of behaviour in social settings with exploration of the meaning these actions have for the participants.

Research Methods
Symbolic interactionism studies both the behaviour of social groups and the meanings which their actions hold for the participants. Those being studied are referred to as 'actors' rather than 'subjects' because their own purposeful activity is taken into account. The researcher's active, interpretive role is also acknowledged. For this reason, great care must be taken when entering the social worlds of others. Their own view of what it is they are up to is part of what a researcher must consider in forming his or her own theory.

It is clear that a study of children and television which follows this method will seek out children's own ways of thinking about the experience of television viewing and relate these to their viewing behaviour. Because this has seldom been done, it was chosen as the beginning step in a three-stage research design, which used a combination of methods to explore children's relationship with television. The findings based on this new theoretical and methodological approach give a very different picture of children and television, and may herald a change in the way television is regarded in the education of children.

Stage 1: Interviews
During September and October 1982, sixty-four children were interviewed at four primary schools in different areas of Sydney, Australia. Two schools were state schools; two were private (Catholic) schools. Children were chosen from class lists to prevent teacher bias. There were equal numbers of boys and girls, and equal numbers from year three (8–9 year olds) and year six (11–12 year olds). Interviews were conducted one-to-one but in an attempt to minimise control of the talk by the adult, a conversational method was used, exploring the words and concepts which children introduced. An overall thematic guide was used which focused on children's definitions of the television world, what they liked and disliked on TV and television's part in relation to friends, family and school. Transcripts of the tapes (700 pages) were read and re-read to establish common themes and relationships between them.

Stage 2: Observations

During October, November and December 1982, twenty-three children were observed at home watching TV and doing other routine activities. Boys and girls, 8–9 year olds and 11–12 year olds, were observed in equal numbers, except for 11–12 year old girls, of whom there were only five. Families were contacted using a variety of networks and methods. This was difficult, and the search for families continued through to the last weeks.

The observations were for a total of nine hours for each child, divided into three sessions of three hours in the afternoon/evening period during one week. The observer (there were a total of four including an anthropologist) kept a continuing account of everything the child said and did during the nine hours, except during dinner if the family ate together. (The observer did not eat dinner with the family.) The accounts were written as ongoing dramatic scripts which came to be called 'diaries'. If the child watched television, a note was made at appropriate times of what was being shown.

Children were introduced to the observer as a person from the university who was interested in what they watched on television. They were asked to do the things they normally did, even if it was not watching TV. In most cases they paid little attention to the observer after the first few hours.

Observation data was supplemented by an interview with the mother, conducted before the observations began, and a diary of before and after school activities filled in by the child for the three days of the observation. At the end of the observation period children also answered a very short questionnaire.

Observer's comments were sought during and after each observation, but were clearly marked and separated from the observation data. Meetings of observers were held each week during the conduct of observations to enable feedback, clarification and comparison of methods. They were also necessary for morale. It was difficult to work in someone else's private space in a sustained way. In meetings, we could let off steam and still maintain the confidentiality of the co-operating families.

Analysis involved a lengthy process of reading diaries, summarising viewing events, then re-reading and developing categories of behaviour. These were listed and then tested in a review of the diaries.

Stage 3: Survey

During July and August 1983, 486 children in fifteen schools in Sydney were surveyed by questionnaire. There were approximately the same number of boys (239) and girls (247), and of 8–9 year olds (248) and 11–12 year olds (238). Children were chosen at random using a slightly modified form of cluster sampling, so that the results could be generalised to Sydney children in the two age groups. Children from all social classes and most ethnic groups were represented.

The questionnaire was mostly hand-printed and was set out to look like an attractive school workbook. Some of the questions were composed of a series of pictures which had to be circled or crossed through. The pictures themselves were based on others drawn by children during the interviews. Usually, the survey was administered in groups of twelve children of the same age, and the researcher read every question and adjusted the pace for each group. The completed forms were coded, punched and tabulated in standard ways. The following discussion does not treat the three research stages separately but integrates the findings into a single report.

Television Viewing as Children Understand It
A knowlege of children's ways of thinking about their television experience was to be the foundation of this research. The first two questions in the interviews showed that children's definition of television viewing was selective and based mainly on their own positive experiences of favourite programmes which they viewed regularly. In response to the questions, 'Do you watch TV?' and 'What do you watch?' all but nine children immediately listed the names of specific programmes. Only cartoons and movies, and for a few of the older children 'science fiction' and series, were ever mentioned as a group. Not one child included in their initial talk about TV those programmes which they disliked or did not watch. Nor did they mention commercials, even though, when they were questioned about them, they expressed very definite opinions.

Their selective definition of television viewing was also clear. Children would often complain, 'nothing's on' when one of their regular favourites was not available, even though they left the TV on and might even watch another programme. In the following excerpt, Debbie's TV guide is not like the one in the newspaper but is a list of the programmes she likes best:

> *Do you usually watch the same things every week?*
> Yeah ... except for the movies.
> *So you always know what's on television?*
> Yeah ... mum says I've got me own TV guide in me brain.

The association of favourite programmes with television viewing means that talk about television will usually evoke very positive feelings. Children's enjoyment of viewing certainly came through in the drawings they did, during the interview, of themselves watching TV. In these drawings there is a visual expression of the words 'fun' and 'excitement' which they used often to describe their reasons for liking particular TV shows. Children consistently drew themselves smiling as they watched TV and if the television screen faced out of the page, a scene from a favourite show would appear. Dramatic moments or family scenes with everyone smiling were common subjects.

The association of fun and excitement with television viewing is not often canvassed in adult discussion or in research. Children themselves are aware of the disparity between their own opinion and that of adults in general. When asked in the survey to rate the opinions of themselves, their mother and their teacher, 59% of the children stated their opinion to be 'TV is great'. By contrast, only 14% said this was their mother's opinion and 12% said this was their teacher's opinion. Children were most likely to attribute to adults the opinion, 'sometimes TV shows are good' (mothers 40% and teachers 45%).

Children's talk about TV in the interviews expressed their positive feelings about viewing; in words, pictures and even in the gestures and sounds they used to describe programmes. They talked about their viewing as if they exercised freedom in the choice of programmes, and were able to be involved one-to-one.

Evidence from the survey did not support the impression that this was the usual social situation for children watching television. Only a small minority of children (7%) said they watched TV alone in the afternoon or in the evening to 9 p.m. (1%).

Both of these aspects of children's viewing were confirmed in the observation study. Here, children were seen trying to create a one-to-one relationship with the TV set whatever the physical or social surroundings. While this could not usually be managed for all of their TV viewing it was especially notable for the one or two shows with which children wanted to be most involved.

The most common way children contrived to be involved with TV was a physical one; they sat or lay down very close to the TV. The sense of cosy intimacy with TV was usually reflected in the physical appearance of the viewing area. Beanbags, cushions, blankets and comfortable lounges were the furniture of child TV viewing. Children usually sat in their familiar spot each day, using soft furnishings to get very comfortable and relaxed. Where the space was shared with siblings, attempts to modify surroundings or grab the best spot for viewing sometimes created conflict. In one family children were observed each day vying for a particular corner lounge suite, then fighting over who had the most blanket.

Distance from the television set was sometimes an indication of how important a show was for children. One boy, Peter, was seen to adjust his distance from the TV set often over the three days' observation, depending on what was on and who else watched with him. On one occasion as he watched *M*A*S*H*, his favourite programme, he was frequently disturbed by the noise of his mother and younger siblings playing together on the same lounge. He gradually moved closer to the TV, turning up the volume each time, and became insensible to the family's activity. His mother apologised when she noticed him perched on the end of the lounge with the volume much louder than usual.

Although children want to be involved with some shows, they don't necessarily want to watch alone. When family members were disinterested or too noisy, pets were sometimes chosen as viewing companions. During the interviews children had surprised interviewers by naming pets as 'people' they watched TV with. Some even described the pet's viewing habits. Their presence was confirmed in the survey where just under half of the children (45%) said they watched TV with a pet. Of these, 16% watched with a dog, 22% with a cat and some with both. Birds, rabbits and even fish were named as viewing companions.

Seven of the twenty-three children in the observation study watched with cats or dogs. Sometimes they would play with them or feed them during programme breaks, but most often both would sit in close physical contact. When children wanted to watch TV closely, the pet's presence seemed to aid concentration by providing something to be with and to touch.

Reading this description of children's physical arrangements around the TV set may prompt some critics to comment that this is the very anti-social behaviour of which television has been accused. On the contrary, the deliberate organisation by children of the little space that is left to them shows a keen sensitivity to their social and physical surroundings. Children were adept at manipulating their own environment to achieve their purpose of close involvement with the human drama on screen. This kind of TV viewing is anti-social only in the very narrow sense of annoying parents or not interacting with others face-to-face for a time. As the following section will demonstrate, it is unlikely that even the second charge can be applied to much of children's viewing.

Social Interaction During Television Viewing
One of the devices of the 'nasty medium' argument is to assume that children's relationship with television is unique to the medium, that it is unlike other kinds of social behaviour. By contrast, even the early work of the symbolic interactionists defined the relationship as akin to everyday social interaction but without the communication going both ways. In their early study of television hosts in game shows Horton and Wohl described the audience's relationship as 'in-

144

timacy at a distance' where television made available the kind of 'nuances of appearance and gesture to which ordinary social perception is attentive and to which interaction is cued'.[11]

Recent research and theoretical writing in this field has confirmed such a view. In their observational study of children watching TV, Reid and Frazer describe instances of children playing with each other around the set as they viewed and even incorporating material from on screen into their talk and actions. In another study, based on similar research methods, these authors demonstrate how children use TV content to initiate social events, making comments to and demands upon those around them, including parents, based on the commercials they were watching at the time.[12]

While this finding does not surprise anyone who has watched television with children, it belies the simple model of children's viewing being something apart from and different to everyday social behaviour. Fine and Kleinman make this general point about interaction with the mass media:

> popular culture products are created in group interaction, require group facilitation for their transmission and frequently are viewed by interacting groups ... Thus, even a communication form seemingly removed from face to face interaction – the mass media – is grounded in the same set of basic interactional criteria.[13]

Designing a new study which would observe children watching TV at home provided the opportunity to analyse the different ways that children interacted with television and with each other as they viewed. Murray has described four different types of viewing behaviour: 'staring', 'commentary' (talk about programmes), 'imitation' (copying actions from TV) and 'reproduction' (acting out a whole sequence from TV).[14] In the new study, eleven categories of interactions were derived from reading and summarising over 200 hours of observations. These categories were given the name 'forms of interaction' and are listed in Table I.

The forms of interaction and their definitions were based on a method of constant comparison between the activity list for each child. At first the categories were too general to reflect adequately the variation in behaviour between children and in different circumstances. As the analysis progressed, concepts were refined and revised so that they were sufficiently powerful to make those distinctions clear.

Two groupings have been created: 'expressive' forms of interaction, where television content is revealed in the child's behaviour, and 'non-expressive' forms where their interaction with television is inferred from children's gestures and the regular association of behaviours with TV events. In each column, forms of interaction are listed in an order which shows children's attention to the TV screen. Parasocial interaction and intent viewing, for example, both involve very close and continuing attention to TV programmes, whereas re-make usually takes place away from the set at another time.

TABLE I

Forms of interaction while watching television

Expressive	Non-expressive
Parasocial Interaction	Intent viewing
Performance	Systematic switching
Comment	Partial distraction
Discussion	Anticipation restlessness
Self-talk	
Monitoring	Monitoring
Re-make	

Four forms of interaction will not be discussed at length in this paper: systemic switching, partial distraction, anticipation rest-lessness and monitoring. In the last three, family events or leisure activities command children's attention as well as television. They therefore give it strategic notice in ways which have also been documented in laboratory studies.[15] Systemic switching is a way of watching two or more programmes during the same interval of time. These forms of interaction confirm the adroitness of children in nego-tiating their viewing alongside other distractions. However, other forms reveal more about children's social relationships as they view.

Parasocial interaction was identified by Horton and Wohl soon after the introduction of television as the 'simulacrum of conversa-tional give and take' in which the viewer 'observes and participates in the show by turns'.[16] This participation may be in response to a person on the show or to other structural elements. This kind of involvement with television did not mean that the child was 'glued' to the TV set and cut off from others. Most often it was a joint exercise with another child, as in the following example where two children are watching *Play School*, an Australian pre-school prog-ramme:

(On the television screen a calendar appears with a blank space for the day)
Emma: Today is Friday, 10th December.
(Later in the programme the presenter talks about ice blocks: *'That's hard and cold isn't it?'*)
Emma: We can't feel them!

146

(The presenter makes a nest with her hand; Emma and her sister
do the same)
(*'Do you ever wear mittens when it is cold?'*)
Emma: Yes.
Sister: Yes.
(The children continue to respond to the actions of the presenters,
in turn rubbing hands, stamping feet and jumping up and down.)

Performance is similar to parasocial interaction, but takes place at
the same time as the TV events, not like a conversation. Advertisements, because they are so familiar to children and often contain
jingles, are likely subjects for performance. In some families, especially those who lived in close proximity and shared the one recreation space with the TV, children would 'perform' with the TV programme, even if they were not sitting watching. Sometimes they were
not even in the TV room, and relied on aural cues for their performance. The television is going at Judy's place but she is not watching,
because she and her mother are busy getting things together to go
away for the weekend. When an advertisement for soup is heard they
each sing different parts:

Mother: Flavour, flavour, flavour.
Judy: Maggi, Maggi, Maggi ... is it maggi or magic?
Mother: Maggi.
Judy: (as she hugs her mother) I love you.

Parasocial interaction and performance were forms of interaction
which showed most obviously children's enjoyment of particular
programmes. If another child or a responsive adult was present, this
pleasure was usually shared.
Discussion and *self-talk*, by definition, involve person-to-person
communication while viewing. What children saw on television
prompted questions or extended talk to those who were viewing with
them, and to parents if they were close by. In self-talk, children not
only discussed television content, but also related it to their own
experience. Phillip sits at a table drawing and listening to the news
on TV. One of the items describes a murder:

Phillip: There was a murder round here not long ago. I didn't know
what happened and I looked over and they were filming it.

Even *comment*, a short pronouncement or 'aside' which children
made as they viewed, often required an 'audience' in the shape of
co-viewers. Two children in the study made short comments to the
cat or dog who watched with them. In the following example, Peggy's

147

initial comment is part of an ongoing discussion with her mother about the events of *Sons and Daughters*, an Australian serial drama.

> (On the screen a couple embrace)
> *Peggy*: He doesn't love her. (She looks to her mother for confirmation);
> (On the screen a car draws up to the house)
> *Peggy*: Mum, would that be filmed at the house?
> *Mother*: Sometimes filmed at the house, sometimes a studio set.
> (The woman on TV decides to keep her child)
> *Peggy*: Yeah! We won the bet!

Peggy and her mother continue discussing the programme after it has finished and during the commercial until the beginning of the next progamme.

Interactions with programmes and with co-viewers in which children use what is on TV in front of them demonstrate that viewing is 'social' in at least two ways. Firstly, children's ways of relating to TV are part of their everyday social behaviour. The children who displayed these forms of interaction in their viewing behaviour did not 'shift gears' and enter into the zombie-like trance described by some writers. Secondly, television viewing was often part of the ongoing social relationships within families. Children made use of whoever was nearby; as joint participants, as a source of information or as a live audience for children's own comments.

There were two occasions when children were observed at play, using what they had previously seen on TV for their own enjoyment. This kind of interaction, which uses remembered television content, has been called 're-make'. It is likely that play using TV is common between friends, but the location of the observation at home limited its occurrence in this study. The most extended example was between Suzi, aged 8, and her friend as they played in her bedroom one afternoon. For a few minutes they listened to a music box which was in the shape of a small colour TV. Then they began to take turns speaking into a cassette recorder and playing it back. They pretended to read the news on TV, making it up as they went: 'Today Tasmania was destroyed but Russia is going to be chopped up to make lots of little Tasmanias' and 'Be careful, because some men are going around Sydney stealing knickers'. Their play was punctuated by much giggling, and sometimes they slipped into American accents. The following afternoon, Suzi spent twenty minutes lying on her bed listening to the recording they had made.

Children's play using television may be 'social' in a more profound sense than those which have already been described. In their play, children have control over what they will choose, how they will perform and the kinds of comments they make in the process. Although there are indications in other areas of this research project that children prefer close imitation of TV events, their own repetition and rehearsal of TV content is a way of deciding with friends what to learn from TV and how to regard it. Suzi's spoof of the news is compatible with the finding that the majority of 8–9 year olds hated news programmes above all others.

There was one form of interaction which all but two children in the observation study engaged in. This has been named *intent viewing* to convey the level of concentrated attention. Intent viewing is part of the viewing behaviour of most children. 'Just watch TV' was the activity circled most frequently by children in the survey, where 85% said that it was one of the ways they usually viewed. However, it was never the only viewing activity they circled. Intent viewing has been given other names in research and in journalistic writing – 'staring', 'passive gazing' and 'zombie-like trance', to name a few. By neglecting to observe the range of children's viewing behaviour, it has been easy to characterise the whole of children's viewing in this way. Intent viewing is defined by close attention to the screen, a relaxed posture and very little bodily movement, except for slight movements of the hands and feet. Children often stroked animals, sucked pencils or played with small objects as they concentrated.

The period of time children spent watching intently varied greatly from child to child. Some would watch a whole programme this way, with brief moments of distraction. Others could only sustain this attention for minutes at a time, usually in the most dramatic segments of a programme. It became very clear during the observations that children were deliberate about what they attended to closely. Intent viewing was never sustained for the duration of the commercial break and, for 11–12 year olds, the change in viewing behaviour when advertisements began was immediate. Intent viewing was usually associated with the programmes children liked best, as was discovered at the end of observations, when children provided the names of their favourite shows.

Len, aged 8, is an example of a child who sustained intent viewing for some time. During Len's favourite programme, *Dr Who*, he watched intently for ten minutes, was distracted during a long discussion of strategy between the characters, then watched intently again for ten minutes. During this time, he sucked his finger and moved to support his head on his hand. Len opened his eyes wider when the main characters were running from danger or engaged in fighting enemies. For the whole of the programme, there were loud noises of children playing just outside the door, which Len ignored. As soon as *Dr Who* was over he went outside to join them.

When children are closely observed watching television, the

149

nuances in their behaviour become clear. Their attention to television is selective and deliberate. Even when they give a TV programme their full attention, they respond in a variety of ways to the social dramas portrayed on the screen. If others are present, the response will probably be acknowledged or shared in some way. It is possible that the children in this study were unusually astute in their management of their social environment and their responsiveness to social interaction on TV. There may be 'zombies' out there that have gone undetected. I am more inclined to think that it is an adult prejudice against the television medium that has produced the 'zombies' in our imagination. If it were not associated with television but with classroom teaching, children's 'intent viewing' would probably be rewarded with praise.

The Social Class Nature of Children's Television Viewing
What you win in discovering the meaning and complexity of children's relationship with television, you lose in terms of the traditional demands of empirical work in this field. The detailed observation of twenty-three children does not establish the links between their interactive viewing behaviour and more conventional measures of social class and amount of viewing which have been the standard indicators. In the original research design a survey had been planned to allow 'measurement' of any 'variables' which emerged as being important in the child/TV relationship. It also reflected the concern within symbolic interactionist writings that research about social process should not ignore the structural relationships established within a particular field.

One of the most consistent findings in surveys about children's television viewing has been the relationship between longer hours of viewing and lower socio-economic status. This relationship still holds good in Australia, according to a 1982 survey of pre-school children.[17] It was confirmed in the present study, where the amount of TV viewing to 9 p.m. was positively correlated ($r = .20$, $n = 486$, $p < .001$) with low occupational status (the father's; many schools did not even record mother's occupation). However, 'the identification of a class-viewing relationship does not necessarily advance our understanding of why some children watch more television than others'.[18] Nor does it address the different ways that children interact with TV at home. The final stage of research therefore provided an opportunity to specify some of the relations between social class or amount of TV viewing and newly-developed measures of children's leisure and their activities while viewing.

The most important and novel of these was the Viewer Activity Index. Derived from interviews and observations it used pictures to illustrate nineteen different activities around the TV set, for example, 'eat a snack', 'read', 'jump and dance', 'do homework'. The illustrations were reminiscent of children's own drawings in their appeal and humour. Children were asked to circle the picture if it was something they usually did and to put a cross through it if it was something they did not do. The measure established for a representative group of children the variety and frequency of TV activities.

Two different measures of children's leisure activity were developed. The Home Recreation Measure also used pictures to describe the things children commonly did around the house and in their neighbourhood, such as 'go shopping', 'listen to the radio', 'play in the park'. For the second, the Structured Leisure Measure, children wrote down for every week day, the special things their parents paid for them to do. The two leisure questions were quite distinctive in the groups of children they described. The measures were not correlated with each other ($r = -.02$, $p = .37$). They also bore an opposite relationship with socio-economic status. Children who engaged in a greater number of activities around the house and neighbourhood were those of lower socio-economic status ($r = .17$, $p < .001$). By contrast, these children wrote down far fewer structured leisure activities ($r = -.20$, $p < .001$).

Children who enjoyed most of their leisure around home were also those watching longer hours of television. Even when the influence of socio-economic status is controlled for (by the use of partial correlation), this relationship holds good. It seems that the lack of provision of special activities which children do away from home is one significant factor in the amount of TV they watch. Of course, families with more money to spend on recreation, and those who live in areas where a range of facilities are usually available, are at an advantage in providing alternatives to television for their children.

The picture of children's viewing revealed by the Viewer Activity Index would come as a surprise to those who still held ideas of children being passive as they viewed. It was found that the children who circled most activities were also those who watched longer hours of TV ($r = .428$, $p < .000$). This result was based on a partial correlation where socio-economic level was taken into account. It seems that in homes where the television is on for long hours, children do many other things around it. They have adapted television to the performance of other tasks, just as they have adapted themselves to the presence of television. Television and the child's leisure activities occur together in the social life of the family.

Conclusion
With much of this information, people will want to argue about what is 'good' for children and what is 'bad'. For example, it is possibly a very useful social skill to learn to adapt to the constant presence of

electronic media. What children 'lose' by doing so is open to debate. Certainly, the supposed virtue of severely limiting the amount of television children watch must be questioned.

Whatever practices adults adopt, they will need to be more aware of the dynamic relationship children have with television and its special significance to those children who have little access to other kinds of leisure. By acknowledging that children have a relationship with the programmes they view, one which affords them much pleasure and which they like to share with others, greater significance is given to the content of the television progammes themselves. It is these programmes, different for different groups of children, which are showing them social behaviour and its consequences.

The programmes children see do not do justice to their viewing abilities or their curiosity about human relationships. We have been disgracefully slow in coming to terms with our own ambivalence about the television medium and as a result we have handed most of the initiative to define 'children's TV' to the commercial sphere. For those working in commercial television, the bottom line is the maximisation of audiences so that they can be sold as markets.[19] Being suitably appalled at some of the programmes which are shown to children on commercial TV is a good excuse to do nothing. It is encouraging to discover that children are so positive about their television viewing experience. We must give them programmes which deserve their enthusiasm.

References

1. See, for example, M. Winn, *The Plug-In Drug* (New York: Penguin, rev. ed., 1985).
2. G. Noble, *Children in Front of the Small Screen* (Beverly Hills: Sage, 1975); C. Feilitzen et al., *Open Your Eyes to Children's Viewing* (Stockholm: Sveriges Radio Audience and Programme Research Department, 1977); C. F. Frazer, 'The Social Character of Children's Television Viewing', in *Communication Research*, vol. 8, 1981; M. Rice, E. Wartella, 'Television as a Medium of Communication: Implications for How to Regard the Child Viewer', in *Journal of Broadcasting*, vol. 25, 1981; J. Bryant, D. R. Anderson (eds.), *Children's Understanding of Television* (New York: Academic Press, 1983).
3. G. H. Mead, *Mind, Self and Society* (Chicago: Univ. Chicago Press, 1934).
4. H. Blumer, *Symbolic Interactionism: Perspective and Method* (New Jersey: Prentice Hall, 1969).
5. R. L. Schmilt, *The Reference Other Orientation* (Carbondale: Southern Illinois Univ. Press, 1972); T. Shibutani, 'Reference Groups and Social Control', in A. M. Rose (ed.), *Social Behaviour and Social Processes* (London: Routledge & Kegan Paul, 1972); J. A. Anderson, 'Research on Children and Television: A Critique', *Journal of Broadcasting*, vol. 25, 1981.
6. A. Strauss, *Negotiations: Varieties, Contexts, Processes and Social Order* (San Francisco: Jossey-Bass, 1978); P. M. Hall, 'Structuring Symbolic Interaction: Communication and Power', in D. Nimmo (ed.), *Communication Yearbook 4* (New Brunswick: Transaction Books, 1980); S. Stryker, *Symbolic Interaction-*

ism: A Social Structural Version (Menlo Park, CA: Benjamin/Cummings, 1980).

7. Blumer, *Symbolic Interactionism*, p. 187.
8. See P. L. Berger, T. Luckmann, *The Social Construction of Reality* (Harmondsworth: Penguin, 1966).
9. C. F. Frazer, 'Social Character [...]'; L. N. Reid, 'Viewing Rules as Mediating Factors of Children's Responses to Commercials', in *Journal of Broadcasting*, vol. 23, 1979.
10. J. Lull, 'How Families Select Television Programmes: A Mass-Observational Study', *Journal of Broadcasting*, vol. 26, 1982.
11. D. Horton, R. R. Wohl, 'Mass Communication and Para-Social Interaction', in *Psychiatry*, vol. 19, 1956.
12. L. N. Reid, C. Frazer, 'Children's Use of Television Commercials to Initiate Social Interaction in Family Viewing Situations', in *Journal of Broadcasting*, vol. 24, 1980.
13. G. A. Fine, S. Kleinman, 'Rethinking Subculture: An Interactionist Analysis', in *American Journal of Sociology*, vol. 85, 1979.
14. J. P. Murray, 'Television in Inner-City Homes: Viewing and Behaviour of Young Boys', in E. A. Rubinstein et al., *Television and Social Behaviour*, vol. 4 (Washington, DC: US Government Printing Office, 1972).
15. Bryant and Anderson, *Children's Understanding of Television*.
16. Horton and Wohl, 'Mass Communication [...]', p. 215.
17. J. Holman, V. A. Braithwaite, 'Parental Lifestyles and Children's Television Viewing', in *Australian Journal of Psychology*, vol. 34, 1982.
18. Ibid., p. 376.
19. W. Melody, *Children's Television: The Economics of Exploitation* (New Haven: Yale Univ. Press, 1973).

L. ROWELL HUESMANN AND RIVA S. BACHRACH

Differential Effects of Television Violence on Kibbutz and City Children

Introduction

A child's perception of the reality of television violence and a child's identification with television characters have been hypothesised to mediate the effects of excessive exposure to violence on the development of aggression in children.[1] Yet the child's social and cultural environments may strongly affect these perceptions. One of the problems in measuring the importance of such societal variables, even in cross-national studies, is that one can seldom find populations of children whose social environment has differed for most of their lives in well prescribed ways. An exception is the population of kibbutz-raised and city-raised children in Israel. There are no *a priori* genetic, physiological or constitutional differences in their populations that should relate to TV habits or aggressive behaviour. Although the child-rearing to which they are exposed, their social environments (particularly in regard to television) and their norms about aggression appear quite different, both populations belong to the same society. A comparison of these populations also ameliorates another difficulty frequently encountered in field studies designed to examine how social factors affect one's perceptions of television and reactions to television. Often within a society there is insufficient variation in the social factors for an adequate exploration of their effects. The populations of kibbutz- and city-raised children in Israel differ substantially in their social structure. Thus, they represent valuable comparison groups for examining how perceptions and responses to television differ across social environments and for examining how the child's perceptions of and reactions to television violence may affect aggressive behaviour differently in the different environments.

Daily Life for the Kibbutz and City Child

The extensive differences between the structure of the lives of kibbutz-raised and city-raised children in Israel have been well documented elsewhere[2] and will only be briefly summarised here.

The city children typically attend school six days a week from 8 a.m. to noon or 1 p.m. The rest of their time is spent at home or in activities with their peers. Since women work in the majority of Israeli homes, the child is often supervised in the afternoon by an older sibling, neighbour or hired caretaker. However, since the main meal of the day is generally taken at noontime in Israel, many working mothers do come home and see the child at that time. In the evening the family will eat a small meal, generally after 7 p.m. when the shops have closed. City schools usually have relatively large classes of up to 40 children, with a substantial turnover from year to year of both children and teachers. While children may form strong bonds with individual peers, there is by no means the general group cohesiveness and tightly knit peer bonds that exist among kibbutz children.

Typically, a kibbutz-raised child will live with mostly the same peers from birth through to the end of high school. The size of the same age peer group or class is likely to be closer to 15 than 40. Since children are constantly taught that one must always consider the good of the group, it is not surprising that a closely bonded peer group emerges in which prosocial behaviour is emphasised. The kibbutz-raised child's daily schedule during the elementary school years will be somewhat different from the city child's. After waking at about 6.30 a.m. the children immediately go to the classroom which is in the same building as the sleeping and dining rooms. After about an hour of instruction, breakfast will be taken. Formal classroom instruction generally ends by noon when the main meal is eaten. Often a break from classes will be taken during the morning for the children to work on kibbutz projects. After the midday meal, the children will spend time in smaller groups on sports, crafts and other extra-curricular activities as well as resting in their sleeping-rooms. At 4.30 p.m. the children go to their parents' quarters where they may eat snacks, do homework, talk and play with their parents, siblings and friends, or watch television. Around 7 p.m. the children eat the evening meal with their parents either in the kibbutz dining hall or in the parents' quarters. After dinner they return to their children's house where they may do homework, read, or have group meetings and activities. Once or twice a week they may watch TV as a group at this time. By 9 p.m. they are going to sleep.

Exposure to Violence
While violence in Israel is an ever-present daily threat for most people, children seldom see violent acts in person. Terrorism and war affect everyone, but more often indirectly rather than directly. Still, crime and concern about crime have generally been increasing in Israel over the last decade.[3] However, the crime rates in kibbutz

155

society remain much lower than in the city environment. For example, between 1969 and 1975 over 34,500 juveniles earned police records. Of these, only 250 or about 0.7% were kibbutz children, though kibbutz children comprise about 3% of the population.[4] According to clinical archives, the rates of cases of psychopathic behaviour are also significantly lower in the kibbutz population than in the city population.[5] Thus, it seems fair to conclude that kibbutz children are exposed to less interpersonal violence in their environment than are city children. In the next section we argue that the same can be said about exposure to violence in the media.

Television Programming in Israel
Television programming in Israel is government controlled and was introduced in 1969. There is only one channel, with a limited selection of programmes (partially due to censorship). About 60 per cent of the programmes are of foreign origin, mostly from the United States and the United Kingdom. Foreign programmes are shown with the original sound track and subtitles in Hebrew. Many Israeli households can also receive Jordanian TV, although, again, this is also censored and contains mostly imported programmes. Furthermore, since the subtitles are in Arabic and most young Israeli children cannot read it, it is doubtful that it has much impact on them.

Israeli television broadcasting is highly structured in terms of scheduling. Throughout the morning educational programmes are broadcast and are intended mostly for children. From noon until 3 p.m. is devoted entirely to the 'open university' programmes. Some days there may be no television at this time. From 3–6.30 p.m. there are mainly children's programmes, which may include educational programmes similar to *Sesame Street*, children's entertainment programmes (including cartoons which may be violent) or adult programmes considered particularly appropriate for children, for example, *Dif'rent Strokes*. From 6.30–8 p.m. Arabic language programmes are broadcast. Any children's shows broadcast during this time have subtitles in Hebrew. Generally, none of these shows will contain dramatic violence. From 8–9 p.m. family-oriented programmes such as quiz shows, documentaries, family dramas (such as *Upstairs Downstairs*), or comedies (such as *Love Boat*) are shown. At 9 p.m. the national news comes on, consisting mostly of visuals of the newscaster and rarely visual portrayals of violence. Of course, in content the news on Israeli television often contains reports of severe violence. After the news, from 9.30 to 10 or 10.30 p.m. the typical programming is documentary, addressing political, scientific or educational issues. Finally, from 10 or 10.30 p.m. until midnight adult dramas are broadcast which may at times include extensive violence – for example, *The Professionals* and *Charlie's Angels* were shown during the course of the study. Dramas with less explicit violence (such as *Dallas*) are often shown at this time too. The broadcasting day closes with the final news shortly after midnight. On Saturdays there is no broadcasting at all and on Friday afternoons none from 3–6.30 p.m.

Television Viewing for Kibbutz and City Children
Exposure to television is quite different for children living in the city and those raised on the kibbutz. The differing structure of their daily lives, as documented above, strongly affects their access to television. While the city child has his or her afternoon relatively free of required activities and therefore can spend that time watching television, the kibbutz child is occupied until the parental visiting time at 4.30 p.m. During the kibbutz child's visit with its parents, the child may watch television. However, when the child returns to his peer group after the evening meal, it would be only rarely that more television would be watched. For the city child, on the other hand, television viewing all evening is possible if his or her parents consent. Since, as documented above, most seriously violent programmes are on late at night, the kibbutz child's access to such programmes is far more limited than the city child's. Furthermore, it is much more likely that the kibbutz child would observe any such programme with a group of his or her peers and the 'metapelet' ('caretaker'). This substantial difference between the potential exposure to television violence of kibbutz and city children coupled with the strong group structure and social bonds of the kibbutz child's society lead to the hypothesis that perceptions of television violence and reactions to television violence will be quite different in kibbutz and city-raised children.

Method

Subjects
The subjects in the city sample were children in the first and third grades at two public schools in Ra'anana, Israel. Ra'anana is a small town in the Sharon District, 15 kilometres north of Tel-Aviv. The population consists mostly of upper middle-class families, and many of the fathers and mothers work in Tel-Aviv. The sample consisted of one class of third-graders in both schools and one class of first-graders in one school. This gave us 39 first-graders (19 boys and 20 girls) and 73 third-graders (37 boys and 36 girls) in the original city sample.

The children in the kibbutz sample were residents of two kibbutzim also located north of Tel-Aviv in the Sharon District (Ma'abarot and Ein-Ha'Horesh). Both of these kibbutzim had been in existence for over 40 years and were established by the same socialist political movement ('Hashomer Hatzair'). The entire first and third grade in both kibbutzim were used, giving us a sample of 38 first-graders (16 boys and 22 girls) and 26 third-graders (12 boys and 14 girls). In addition, in one of the kibbutzim 10 second-graders were also studied (4 boys and 6 girls) because they were in the same class with the third-graders. It is common to split classes in kibbutz schools owing to the small number of children in each settlement. Thus, a total of 186 children were studied in the original sample – 74 from kibbutz schools and 112 from city schools.

During the course of the study a number of children dropped out of the city sample; however, only one child (a girl) was lost from the kibbutz sample. At the same time, because it was easier to test entire classes, a substantial number of new subjects were added to the city sample in years two and three. The first-grade class in the city school was broken up after the first year, and its students were redistributed into two second-grade classes. Some were transferred to another school where they could not be tested. This was the greatest single cause of subject attrition in the city sample. For most analyses only the subjects who were present for all three years will be used. Out of the 256 subjects on whom some data were obtained, 158 were interviewed in all three waves. This constituted 85% of the original sample of 186; so the overall mortality rate was only 15%. However, among city children the mortality rate was 24% compared to 1% among kibbutz children.

Procedure
The children were interviewed in the spring three times at one year intervals, giving us an overlapping longitudinal design with data on first to fifth-graders. The younger cohort was in the third grade in the final year (third) of the study while the older cohort was in the third grade in the first year of the study. Each year the children were interviewed in two group sessions lasting about one hour each. In addition, some information was obtained about each child from school records and teacher interviews.

Measures
A child's TV violence viewing score was based on the child's self-report of the shows he/she watched most often. Three lists of nine (1981), seven (1982) and six (1983) programmes each were presented to a child. From each list the child chose the *two* programmes he/she watched most often and then rated how often he/she watched them on a three-point scale. Each list contained popular violent and non-violent programmes, child or adult oriented. The violence of each show was rated by five independent raters (inter-rater reliability

158

was above .70), and their average rating was the programme's score. Two violence viewing scores were computed for each child. One was the simple mean of the violence ratings for the six programmes the child selected. This is called the 'violence of the child's favourite programmes'. For the other score each violence rating was weighted by the frequency of viewing that show as reported by the child. This score is called the child's overall 'TV violence viewing'. A separate summed frequency of viewing score was also computed representing how 'regularly' the child watched his/her favourite shows.

Besides measuring violence viewing and frequency we asked each child questions about his/her belief in the 'realism of certain violent television programmes' and questions about the extent to which the child 'identified with various TV characters'.[6] For each of six violent shows, the child was asked 'how much does the programme tell about life like it is?' 'Does it tell about life just like it is, a little bit like it is, or not at all like it is?' A child who scored high would be one who believed that violent shows tell about life just like it is. The children were also asked the extent to which they thought they resembled (in behaviour) four popular characters: an aggressive male, an aggressive female, an unaggressive male, and an unaggressive female. Two scores were derived – a total identification score, and an identification with aggressive character score.

The primary measure of overt aggression was a modified version of the widely used peer-nominated index of aggression in which each subject in the sample names all the subjects in his class who have displayed ten specific aggressive types of behaviour during the last school year, for example, 'Who starts a fight over nothing?'[7] This measure has been used in many countries and is both highly reliable (Coefficient Alpha > .95) and valid. In the final wave of the study, the children also completed self-ratings of aggression which provided concurrent validation in the Israeli population.

Peer-nominations were also obtained on two questions that measured the children's popularity – 'Whom would you like to sit next to?' and 'Whom would you like to have as a best friend?' One other peer-nomination question was asked (which is relevant for this paper) – 'Who never fights even when picked upon?' This avoidance of aggression question correlated very negatively with the aggression questions. It was found that the ratio of the aggression to the avoidance of aggression scores provided a measure of aggression that corrected for general nomination biases in the two populations.

Two additional measures of the children's behaviour were obtained that were hypothesised to differ across the populations and to be related to TV habits. The child's use of aggressive fantasy was evaluated with part of the Children's Fantasy Inventory.[8] The child answers a number of questions about how often he/she daydreams

159

about or imagines acting in certain ways. From these self-ratings a scale of 'aggressive fantasy' was derived. A child who scores high would be one who frequently has aggressive daydreams. The children's preferences for sex-typed activities were also evaluated with a technique described elsewhere.[9] Over several trials each child selected the games or activities he or she liked best from sets of male, female and neutral activities (as determined by surveys in the society). A subject received three scores representing his or her preference for male, female and neutral activities.

To evaluate the role of possible social class differences between the kibbutz and city samples, measures were taken in a teacher's interview and from school records as to ethnic background, place of birth, parents' education (number of years of schooling) and parents' occupation which was scored for status according to Hartman.[10] Finally, each child completed a Draw-a-Person which was scored for intellectual development.[11]

Results

Before examining the TV habits of the kibbutz and city samples, let us consider how they differ on demographic variables. The differential attrition rate in the two samples is in itself an important demographic difference. This indicates the rather closed and static nature of kibbutz life in which low mobility is the norm. In a number of studies in western countries, higher family mobility has been associated with higher levels of child aggression.[12] Thus, attrition in the city sample might have selectively eliminated the higher aggression children, but, of course, not in the kibbutz sample where there was no attrition.

The ethnic background of the kibbutz and city children also differs significantly. The major difference is that the parents of kibbutz children were more likely to be native born ('sabras') than were the parents of city children. Since ethnic background was only measured in terms of the parents' origins, there is no way of knowing the original ethnicity of the Israeli-born parents.

The number of years of education of the parents in the two samples was about the same: both mothers and fathers averaged about one and one-half years of education beyond high school. Similarly, kibbutz and city fathers did not differ in their average occupational status as measured by the Hartman Occupational Status Scale. However, the standard deviation of occupational status scores was about 21 per cent higher for the city fathers. While there were a few city fathers who were doctors and lawyers and a few with low-status occupations, the statuses of most kibbutz fathers' occupations were homogeneous and in the middle range.

160

The one substantial difference between kibbutz and city families on traditional socioeconomic variables was that kibbutz mothers all work and on the average have lower-status occupations than working city mothers (F(1,109) = 6.4, p < .013). Only 64% of city mothers worked.

In conclusion, although both populations are apparently within the upper-middle range of socioeconomic status, as represented by level of occupation and years of education, the kibbutz group is far more homogeneous than the city group.

Television Habits in Kibbutz and City Children
The mean scores on the major television variables are compared for kibbutz and city children in Table I. City children on the average watched their favourite programmes more regularly (F(1,139) = 12.36, p < .001), were exposed to more television violence (frequency weighted violence, F(1,139) = 4.13, p < .05), and perceived television violence as marginally more like real life than did kibbutz children (F(1,137) = 3.33, p < .07). City children also identified more with TV characters in general (F(1,94) = 5.40, p < .025), and city boys identified more with violent characters (Interaction F(1,108) = 5.14, p < .05). However, kibbutz children's preferred shows were just as violent (F(1,139) = 0.10, n.s.). All these findings are consistent with the picture of kibbutz children living in an environment with far fewer

TABLE I

Television habits in kibbutz and city children

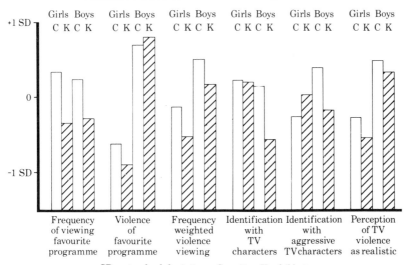

SD – standard deviation   C – city   K – kibbutz

161

opportunities to observe media violence but being similar to city children in their reaction to television.

Demographic Factors and TV Viewing

TV violence viewing did not vary significantly with the family's ethnic origin. However, the same reservations apply to the interpretation of this result – namely ethnicity is much too confounded with kibbutz/city status to permit a clear interpretation.

TV viewing habits do relate significantly to parents' educational status among both populations of children but in different ways, as Table II illustrates. For city boys, frequency of viewing favourite programmes and violence viewing are inversely related to parents' education (r = $-.42$, p < .01 and r = $-.22$, n.s. respectively). This result is comparable to what has been reported from other countries. However, unlike some other countries, the sons of working mothers in the city sample watched less TV (r = $-.38$, p < .05) and less TV violence (r = $-.37$, p < .05) than the sons of non-working mothers. Again, this latter result may be because in the suburban community in which the study was conducted more highly educated mothers are more likely to be working (r = .52, p < .001). For city girls and the kibbutz children, the relations between TV-viewing habits and parental education were quite different than those for city boys. More educated parents had children who watched their favourite

TABLE II

*Correlations of parents' education with TV habits
(averaged over three years)*

| | Parents' education | | | |
| | Girls | | Boys | |
Television viewing variables	City	Kibbutz	City	Kibbutz
Regularity of viewing favourite programmes	.26 (N = 33)	—	$-.42$** (N = 40)	.33† (N = 28)
Violence of favourite programmes	.37* (N = 33)	—	—	.26 (N = 28)
TV violence viewing (regularity × violence)	.47** (N = 33)	.23 (N = 40)	$-.22$ (N = 40)	.33† (N = 28)
Identification with all characters	—	—	$-.25$ (N = 34)	.36 (N = 17)
Identification with violent characters	—	—	—	.56** (N = 22)
Perceived realism of violent programmes	.22 (N = 32)	—	—	.33† (N = 28)

† p < .10   * p < .05   ** p < .01   *** p < .001

162

programmes more regularly and watched more TV violence. This effect was strongest for city girls. However, there were significant effects for both kibbutz boys and kibbutz girls as well. Since most of the kibbutz child's TV viewing is done in the parents' quarters, it is not surprising that these parent variables affect the kibbutz child's TV habits. However, the direction of the effect is surprising. Perhaps the more highly educated and higher status parents spend less time with their children during the 'children's hour'.

Intercorrelations of TV Behaviours

The extent to which regularity of viewing predicts other TV scores is shown in Table III. Among most children regularity is correlated with the perception that TV violence reflects real life. Among girls regularity of viewing is also correlated with a preference for violent programmes, and among city boys it is correlated with greater identification with TV characters.

TABLE III

Correlations of regularity of viewing favourite programmes with other TV habits (averaged over three years)

| | Regularity of viewing | | | |
| | Girls | | Boys | |
Television viewing variables	*City*	*Kibbutz*	*City*	*Kibbutz*
Violence of favourite programmes	.30† (N = 34)	.32* (N = 41)	—	—
TV violence viewing (regularity × violence)	.68*** (N = 34)	.74*** (N = 41)	.75*** (N = 42)	.86*** (N = 30)
Identification with all characters	—	—	.37* (N = 35)	—
Identification with violent characters	—	—	.37* (N = 36)	—
Perceived realism of violent programmes	—	.49** (N = 39)	.25† (N = 42)	.39* (N = 30)

† p < .10 * p < .05 ** p < .01 *** p < .001

What would happen to the kibbutz/city differences in TV habits and perceptions if one controlled statistically the differences in viewing regularity? This was checked with an analysis of covariance using regularity as the covariate (see Table IV). Kibbutz/city differences in the children's perception of how 'true to life' TV violence seems disappeared. However, city children and city boys in particular still identified more with TV characters than did kibbutz boys.

To summarise, kibbutz children watch less television. As a result

TABLE IV
*Results of analyses of variance and covariance of TV habits
with TV regularity as covariate*

Dependent variable	Effect	F Without covariate	F With 'regularity of TV viewing' as covariate
Regularity of viewing favourite programmes	Sex	—	
	Kibbutz	12.36***	
	S × K	—	
Violence of favourite programmes	Sex	80.88***	81.98***
	Kibbutz	—	—
	S × K	—	—
TV violence viewing (regularity × violence)	Sex	20.64***	38.55***
	Kibbutz	4.13*	—
	S × K	—	—
Identification with all characters	Sex	5.09*	4.14*
	Kibbutz	5.40*	3.80*
	S × K	—	—
Identification with violent characters	Sex	—	—
	Kibbutz	—	—
	S × K	5.14†	4.49*
Perceived realism of violent programmes	Sex	19.59***	17.38***
	Kibbutz	3.33†	—
	S × K	—	—

† p < .10   * p < .05   ** p < .01   *** p < .001

they see less television violence, and they are less likely to believe that the violence they see on TV reflects real life. Independently of how much TV they see, kibbutz children are less likely to identify with the characters they do see. Unlike city boys and boys in most other countries, kibbutz boys whose parents are more educated are more likely to watch some violent shows.

TV Habits and Differences in Behaviour
Let us now examine whether these differences in TV habits and perceptions of TV may be related to differences in behaviours that have been observed between kibbutz and city children. One cannot directly compare the kibbutz and city children on the peer-nominated aggression scale because different nomination criteria may have been used in the two samples. However, later we will examine whether peer-nominated aggression relates differentially to TV habits in the two populations. First, though, let us examine two

self-rated characteristics, related to aggression, that do differ between kibbutz and city-raised children – aggressive fantasising and preferred type of games and activities. Table V shows the observed scores on our aggressive fantasy scale for kibbutz and city children. City children (particularly boys) engage in significantly more aggressive fantasy. In Table VI the average scores on the measures of preference for sex-typed activities are shown. Again the differences were significant. Kibbutz children are more likely to prefer less rigidly sex-typed activities (neutral) than are city children. Both of these variables are closely related to aggressive behaviour as Table VII reveals.

Table VIII shows the relation of aggressive fantasy to the child's TV habits and perceptions. For most subjects identification with TV characters is correlated with greater aggressive fantasy. Among kibbutz girls and city boys aggressive fantasies are also correlated with more regular viewing, more violence viewing, and a perception that the violence is real. Kibbutz children may view less TV, but, even so, their individual differences in fantasy behaviour are related to their TV habits and perceptions of TV. Similarly, as Table IX shows, individual differences in preference for sex-typed activities correlates with TV habits and perceptions. Among kibbutz children and city boys, a more masculine and less neutral orientation is associated with higher scores on the TV variables. Among city girls a less femin-

TABLE V

Aggressive fantasy scale for kibbutz and city children
(averaged over three years)

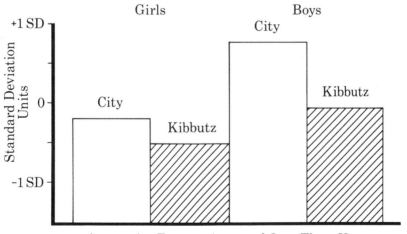

Aggressive Fantasy Averaged Over Three Years

165

TABLE VI

*Measure of preference for sex-typed activities in kibbutz and city children
(averaged over three years)*

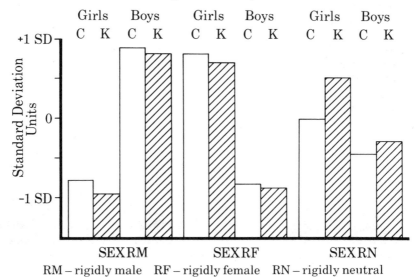

RM – rigidly male   RF – rigidly female   RN – rigidly neutral

TABLE VII

*Correlations of a child's aggressive behaviour with the child's aggressive
fantasy and preference for sex-typed activity (averaged over three years)*

| | Aggressive behaviour | | | |
| | Girls | | Boys | |
	City	Kibbutz	City	Kibbutz
Aggressive fantasy	—	.25	.24	.52**
		(N = 39)	(N = 42)	(N = 29)
Preference for male	.24	.43**	.19	—
sex-typed activity	(N = 35)	(N = 40)	(N = 43)	
Preference for neutral	.27	−.45**	−.28†	—
sex-typed activity	(N = 35)	(N = 40)	(N = 43)	

† p < .10   * p < .05   ** p < .01   *** p < .001

ine (and therefore more masculine or neutral) orientation seems to
be associated with higher TV scores.

These within-sample relations are consistent with the between-
sample differences. Higher scores on the TV variables are associated
with more aggressive fantasy and stronger sex-typing. The city chil-

166

TABLE VIII

*Correlations of a child's aggressive fantasy with the child's TV habits
(averaged over three years)*

| | Aggressive fantasy | | | |
| | Girls | | Boys | |
Television viewing variables	City	Kibbutz	City	Kibbutz
Regularity of viewing	—	.39**	.36**	—
favourite programmes		(N = 40)	(N = 41)	
Violence of favourite	—	.29†	—	—
programmes		(N = 40)		
TV violence viewing	—	.39**	.21	—
(regularity × violence)		(N = 40)	(N = 41)	
Identification with all	.33†	.25	—	.26
characters	(N = 33)	(N = 16)		(N = 18)
Identification with violent	.38*	.38†	.26	—
characters	(N = 31)	(N = 25)	(N = 36)	
Perceived realism of	—	.38*	.36*	—
violent programmes		(N = 39)	(N = 42)	

† p < .10   * p < .05   ** p < .01   *** p < .001

dren score higher on most TV variables and score higher on aggressive fantasy and preference for sex-typed activities. However, when we turn to the relation between aggressive behaviour and TV habits and perceptions, we see some quite different relations within the two populations.

Television Habits and Aggression
Table X contains the correlations between the ratio of aggression nominations to avoidance of aggression nominations with the various television variables averaged over the three years of the study for each gender and population group. The results are very clear cut. Television violence viewing is very significantly correlated with aggressiveness among both city boys and girls but not among kibbutz children. These significant correlations among city children are quite substantial, in fact, much higher than what has been found in most other countries. Interestingly, among kibbutz children, there is not the least indication of a correlation between aggression and violence viewing, but those boys (to a slight extent) and those girls (to a great extent) who perceive themselves as being like TV characters are the more aggressive children. The relation between TV violence viewing and aggression among city children is illustrated in Table XI. City children were divided into the lower 25%, middle 50%, and

TABLE IX

Correlations of preference for sex-typed activities with TV habits (averaged over three years)

	Kind of sex-typed activity							
	Male				Neutral			
	Girls		Boys		Girls		Boys	
Television viewing variables	*City*	*Kibbutz*	*City*	*Kibbutz*	*City*	*Kibbutz*	*City*	*Kibbutz*
Regularity of viewing favourite programmes	.38* (N = 34)	—	—	.49** (N = 30)	—	—	—	-.30† (N = 30)
Violence of favourite programmes	.30† (N = 34)	—	—	—	.43** (N = 34)	—	—	—
TV violence viewing (regularity × violence)	.32† (N = 34)	—	—	.46** (N = 30)	.35* (N = 34)	—	—	—
Identification with all characters	—	—	—	—	—	-.55* (N = 16)	—	—
Identification with violent characters	—	—	.38* (N = 37)	—	—	.55*** (N = 25)	-.38* (N = 37)	—
Perceived realism of violent programmes	.44* (N = 33)	—	—	.30† (N = 30)	—	—	-.30* (N = 43)	-.31† (N = 23)

† p < .10   * p < .05   ** p < .01   *** p < .001

TABLE X

The correlations of average aggressiveness over all three waves with average television viewing

| Television viewing variables | Correlations with aggression | | | |
| | Girls | | Boys | |
	City	Kibbutz	City	Kibbutz
Regularity of viewing favourite programmes	.28 (N = 34)	—	.42** (N = 42)	—
Violence of favourite programmes	.42** (N = 34)	—	.24 (N = 42)	—
TV violence viewing (regularity × violence)	.48** (N = 34)	—	.45** (N = 42)	—
Identification with all characters	—	.66** (N = 15)	.34* (N = 35)	.33 (N = 17)
Identification with violent characters	—	.68*** (N = 24)	.29† (N = 37)	—
Perceived realism of violent programmes	—	—	—	—

† p < .10   * p < .05   ** p < .01   *** p < .001

Aggression was measured as the ratio of peer-nominations on aggressive items to nominations on the aggression avoidance items.

TABLE XI

Relationship between TV violence viewing and aggression among city children over three waves

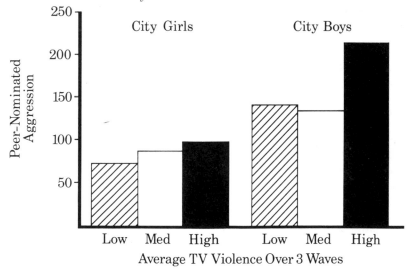

169

upper 25% on TV violence viewing, and their mean aggression scores were plotted.

Before examining the causal ordering of aggression and TV habits with longitudinal analyses, one must ask if the observed correlations might be spurious and due to the effect of a third variable. In particular, can the parent's education, socioeconomic status, the child's IQ or the child's age be generating the relation? In the current study, there were no significant correlations between the child's IQ and either TV habits or aggression. Unfortunately, the IQ scores had to be derived from Draw-a-Person tests given in group settings. Thus, the IQs have questionable validity, and their failure to correlate with either TV habits or aggression should not be given much credence. In most other countries and many previous studies IQ has consistently correlated with both aggression and TV habits. As discussed earlier, parent's education correlated positively with violence viewing and aggression among city girls, negatively among city boys, and not at all among kibbutz children. Father's occupational status displayed similar relations. The child's age correlated differentially with violence viewing depending on the population. Older kibbutz children view TV violence less regularly perhaps because their structured life allows less time for it. Despite these relations, when parent's education, father's occupational status, child's IQ, and child's age were all partialled out of the relation between violence viewing and aggression among city children, these latter variables remained very significantly correlated (.39 for girls and .42 for boys). Thus, the relation between violence viewing and aggression cannot be attributed to these variables.

The causal relation between violence viewing and aggression in the city population can best be examined by the multiple regressions shown in Table XII. It can be seen that a city child's average TV violence viewing over the first two waves of the study was a significant predictor of that child's aggression in the last year. This was true even when initial levels of aggression and avoidance of aggression were partialled out by adding those variables to the regressions, and it was true for both boys and girls. In other words, greater violence viewing was predictive of greater later aggressiveness independently of initial aggression.

Having established that for city but not kibbutz children television violence viewing seems to engender aggressive behaviour, one must ask what factors produce the difference between these populations and what these factors suggest about the psychological processes that are involved. The most obvious differences, of course, are that city children have more control over their TV viewing, watch more TV violence, and perceive the violence as being more like real life. Most violence, in fact, is shown after 9 p.m. when kibbutz children

170

TABLE XII

Multiple regressions predicting city children's aggression
from earlier TV violence viewing
controlling for initial levels of aggression and aggression avoidance

Criterion variable	Predictor variable	Standardised regression coefficients	
		City girls	City boys
Ratio of aggression to aggression avoidance in third wave			
Grade		$-.21$	$-.31^*$
Peer-aggression wave 1		$.55^{**}$	$.58^{***}$
Peer-aggression avoidance wave 1		$-.08$	$-.03$
TV violence viewing waves 1 & 2		$.52^{**}$	$.29^*$
		$R^2 = .46,$	$R^2 = .44,$
		$F_{(4,29)} = 6.13,$	$F_{(4,37)} = 7.3,$
		$p < .001$	$p < .001$

$^\dagger p < .10$ $^* p < .05$ $^{**} p < .01$ $^{***} p < .001$

have few opportunities to observe it. This time factor may also partially explain why the correlations within the city sample are more substantial than those found in other countries. The children who see more violence must be the ones who are staying up later; so violence viewing is confounded with age, parental attitudes and child-rearing practices. Nevertheless, one cannot simply dismiss the effect as an artifact of child's age and parental attitudes. The partial correlations controlling for age and parents' socioeconomic status were not much lower than raw correlations. Kibbutz children also identified less with TV characters than did city children. However, identification with characters related just as strongly to aggression within the kibbutz sample as the city sample. This result is consistent with the data from several other countries as well.

In their analysis of the effects of media violence on children's aggression in Finland and the United States, Huesmann and his colleagues have emphasised the role of social norms for aggression as exacerbating or mitigating factors.[13] For example, they suggest that the reason why females are now susceptible to media violence in the USA but were not twenty years ago is because of the current emphasis on female assertiveness and aggressiveness in the USA. The data on the relation between popularity and aggression in Israel suggests that a similar difference in attitudes about aggression between kibbutz and city children may be contributing to the differential effect of media violence. Aggression was an unpopular behaviour among kibbutz children ($r = -.61$, $p < .001$ for kibbutz boys; $r = -.44$, $p < .01$ for kibbutz girls) but was not related to popularity among city children. This suggests that the imitation of specific interpersonal aggressive acts observed on a TV show or in a movie may be met with far more reprobation when committed by a kibbutz child than when committed by a city child. Since kibbutz children believe that the violence shown on TV is not representative of the real world, and, since they identify less with TV characters (perhaps because they see less TV), they are less likely to accept aggressive behaviour as the norm. Thus, the combination of children having substantially decreased opportunities to observe violence, coupled with peer-attitudes that are less tolerant of aggression and more skeptical about the reality of TV violence, makes it less likely that the aggressive strategies observed on TV will be encoded for later use and retrieved when a child is faced with a social problem.

Summary and Discussion
The subjects in this study came from two distinctly different cultural environments – city and kibbutz – with quite different opportunities to observe TV. In fact, the children's TV habits and perceptions of TV were quite different in the two populations. These TV habits and perceptions were related significantly to the children's preference for sex-typed activities and aggressive fantasising within both populations. Moreover, the between population differences on these variables were consistent with the between-population differences in TV viewing. However, the most notable difference between the two populations involved the relations between TV violence viewing and aggressive behaviour. Among city children a significant positive relation was found between television violence viewing and amount of aggressive behaviour. In fact, the magnitude of the correlation was higher for Israeli city children than for children in any other country. Furthermore, longitudinal effects seemed to be more from violence viewing to aggression than from aggression to violence viewing. City children who viewed violence more often appeared to become more aggressive relative to the rest of the children over the course of the study. To these authors it is surprising that such results should occur in a country in which only a few violent programmes are broadcast late at night each week and in which the environment contains regular examples of real salient violence to which the child is exposed. However, the most important finding in this study is that these relations between exposure to media violence and aggression were not obtained for kibbutz children. More aggressive kibbutz children do identify more with aggressive TV characters, but there is no detectable relation between amount of exposure to TV or TV violence and aggression in kibbutz children.

What are the essential factors associated with growing up on a kibbutz that mitigate the effect of TV violence viewing on the development of aggressive behaviour? The most obvious and direct factor is the different patterns of TV viewing observed among kibbutz children. Because of their daily schedule, kibbutz children are exposed less to television, particularly during the evening hours. This fact minimises their opportunities to watch violent programmes since these (with a few exceptions; for example, cartoons) are broadcast late at night. Like city children and children in other countries, they 'like' action shows containing violence, but these programmes are rarely accessible for kibbutz youngsters.

Another characteristic of kibbutz TV and film exposure that may mitigate against any behavioural effects is group viewing. A number of studies have suggested that the presence of co-observers, and particularly adult co-observers who comment on the material, may mitigate against imitation of the material by the child.[14] The kibbutz child is most likely to watch television with other people. They either watch in the family quarters during the 'afternoon family hour' when the parents are always there, or they watch in the evening with their peer-group and adult 'caretaker'. Moreover, sitting alone in front of a TV would generally not be accepted as an appropriate behaviour for a kibbutz child. Associated with the kibbutz style of life is a continuous stream of organised group activities for children; especially for older children, there are always alternative modes of entertainment to sitting in front of a TV. Thus, it is not surprising that for kibbutz children in Israel, unlike in some other countries, TV viewing declines with age.

In addition to the natural constraints on TV viewing, though, there are other powerful characteristics of the kibbutz environment that probably minimise the impact of TV violence on the development of aggression. Most important, perhaps, is the power of the peer-group. From infancy onward the peer-group prescribes acceptable norms of behaviour and establishes intra-group sanctions against deviations from these norms. For children, peer-group prescriptions for certain pro-social behaviours and sanctions against aggressive behaviour are undoubtedly more powerful factors in social learning than infrequent exposures to television models. Of course, the peer-group's prescriptions are moulded by the adult community of the kibbutz through the 'caretakers' and teachers. However, even the parent's views and behaviours seem to be less important to the kibbutz child than the collective norms. Thus, several parental and demographic variables that were discovered to be related to aggressiveness in city children (ethnic background, parents' education and occupations) were not related to aggressiveness in kibbutz children. At the same time, individual differences in the patterns of TV viewing of kibbutz children were correlated with individual differences in their parents' patterns. Such a result is to be expected since almost the only opportunity that a kibbutz child has for differential viewing is at home during the family hour.

It also seems clear that aggressiveness in general is less acceptable among kibbutz children than city children. Unpopularity was very significantly related to aggressiveness among kibbutz children but not among city children. Of the upper quartile of children who were most aggressive in the kibbutz sample, not one scored in the upper quartile on popularity. Yet of the city children in the upper quartile on aggression, 18% scored in the upper quartile on popularity. Kibbutz children also fantasise less about being aggressive. In short, the powerful group norms against aggression prevalent in the kibbutz society seem to be the most important factor in mitigating the effect of what little media violence the kibbutz child observes. The power of cultural norms to influence aggression has been documented in a number of countries and cultures.[15]

In conclusion, in a children's society in which values and norms of behaviour are clear, where accountability to the society is emphasised, where interpersonal aggression is explicitly criticised, and where solo TV and film viewing are infrequent, the children's aggressive behaviour is influenced very little, if at all, by what media violence they do observe.

The research described in this essay was supported partly by grant MH-38683 and MH-31886 to the senior author.

References

1. L. R. Huesmann et al., 'Intervening Variables in the Television Violence–Aggression Relation: Evidence from Two Countries', in *Developmental Psychology*, vol. 20, 1984.
2. M. Kaffman, 'A Comparison of Psychopathology: Israeli Children from Kibbutz and from Urban Surroundings', in *American Journal of Orthopsychiatry*, vol. 35 no. 3, 1965; A. I. Rabin and B. Beit-Hallami, *Twenty Years Later: Kibbutz Children Grown Up* (New York: Springer, 1982).
3. S. F. Landau and B. Beit-Hallami, 'Israel: Aggression in Psychohistorical Perspective', in A. P. Goldstein, M. H. Segall (eds.), *Aggression in Global Perspective* (New York: Pergamon, 1982).
4. S. F. Landau, personal communication, March 1984.
5. Kaffman, 'Comparison of Psychopathology'.
6. Huesmann et al., 'Intervening Variables'.
7. L. D. Eron et al., *Learning of Aggression in Children* (Boston: Little-Brown, 1971); Huesmann et al., 'Intervening Variables'.
8. E. Rosenfeld et al., 'Measuring Patterns of Fantasy Behavior in Children', in *Journal of Personality and Social Psychology*, vol. 42, 1982.
9. Huesmann et al., 'Intervening Variables'.
10. M. Hartman, *Occupation as a Parameter of Social Status in Israeli Society* (Tel Aviv University: Institute for Research on Work and Society, 1975).

11. D. B. Harris, *Children's Drawings as Measures of Intellectual Maturity: A Revision and Extension of the Goodenough Draw-a-Man Test* (New York: Harcourt, Brace and World, 1963).

12. L. D. Eron et al., 'Aggression and Its Correlates Over 22 Years', in D. H. Crowell et al. (eds.), *Childhood Aggression and Violence: Sources of Influence Prevention and Control* (New York: Academic Press, forthcoming).

13. Huesmann et al., 'Intervening Variables'.

14. H. Adoni, A. Cohen, 'Children's Responses to Televised War News Films', in *Megamot Behavioural Sciences Quarterly*, vol. 25 no. 1, 1979; D. J. Hicks, 'Effects of Co-observer's Sanctions and Adult Presence on Imitative Aggression', in *Child Development*, vol. 35, 1968.

15. S. F. Landau, 'Trends in Violence and Aggression: A Cross Cultural Analysis', in *International Journal of Comparative Sociology*, vol. 25, 1982.

V EFFECTS

IAN TAYLOR

Violence and Video:
For a Social Democratic Perspective

The rapid recent development of the videotape recorder, and videotaped films rented for home use, is provoking considerable attention in many western societies. Advertisers and television professionals alike are obviously concerned as to the likely effects of videotape on the size of their markets and audiences. Other interests, including some parents' associations, educationalists and church people, on the other hand, are anxious over the content of many videotape films, and, in particular, over the relatively easy access to violent and explicitly sexual material that the cheap rental market may afford children and youth.

In Canada, the pressures which have been brought to bear have been such as to promote one of the first examples of coordinated action between provincial governments for some time: in September 1985, a tri-provincial agreement between the governments of Manitoba, Ontario and Saskatchewan signalled the simultaneous adoption in those provinces of the international classification code for videotape films. In 1984, in Ontario, the powers of the Ontario Censor Board had already been broadened by the Ontario Theatres Amendment Act to include the power to regulate, censor, and/or reject, videotape intended for public retail or rental, using the same standards of classification applied to cinema film. Similar developments are pending in the United States, and legislation introducing censorship of videotape material has been introduced in Australia and New Zealand.[1]

The level of anxiety over videotape use and content does seem to be higher in the UK, however, than in other societies. There are at least two primary reasons. One is that VCR use is significantly more widespread in the UK than in any other western society. One estimate for 1983 was that 30.1% of all TV-owners in Britain had a VCR; by contrast, only 10.7% of TV-owners in the USA and 8.4% in Canada were estimated to have a VCR.[2] At that time (1983), 6 million of the world's 36 million VCRs (17.9%) were in British homes.[3] By January 1986, the estimate was that 40% of adults in Britain had a

VCR in their household, some 79% of which were owned.[4] Video-rental stores are now very much a feature of the British urban landscape.

But the second reason for the high level of attention being given to videotape use and content in Britain is the intensive and successful campaign that has been waged by various 'moral entrepreneurial' groups, pressuring the government for tighter regulation and censorship both of videotape, and, in similar legislation, television films in general.

In late 1985 Mr Winston Churchill MP introduced into the House of Commons the Obscene Publications (Protection of Children, etc.) Amendment Bill seeking to extend the provision of the Obscene Publications Act of 1959 and 1964 to television and sound broadcasting. It also, crucially, proposed a new test of obscenity by providing that any article portraying certain activities shall be deemed *per se* obscene. If these scenes are contained in a film, and also if it is accessible to persons under 18 or is broadcast, the Bill proposes that the famous 'deprave and corrupt' test will be inapplicable.

We shall recite some of the recent history of intensified State regulation of visual materials of a sexual or violent character below, in order to provide some grounded sense of the direction of government policy. For our particular purposes, however, we should recall that one of the key moments in the escalation of public, political debate – in the press and the media generally – was the publication, in November 1983, of the report *Video Violence and Children*.[5] This report was the result of a questionnaire survey conducted with 6,000–7,000 schoolchildren, enquiring into their viewing of videotapes currently available in videostores, and especially some 30 so-called 'video nasties' (featuring extreme scenes of violence and violent sexuality). The main findings of this report, issued under the name of the Parliamentary Video Group Enquiry (PVGE), were given headline treatment by the media.

The report has since been very substantially discredited on the grounds, in particular, of the indefensible, idiosyncratic and premature interpretation of data exhibited by Clifford Hill, the prime author and moving force behind the PVGE, but this has not prevented *Video Violence and Children* from having an enormous influence on public, political debate both in Britain and overseas. The publication, in 1985, of a rather more extended and careful analysis of the PVGE questionnaire data, under the same title but with several, quite disparate authors and commentators, has done little to correct the misleading impressions created by the original PVGE Report.[6]

Socialist Responses to the Moral Panic Over Sex and Violence
The truth is, of course, that Britain in the period between 1982 and

178

1985 was in the middle of one of its most intense periods of moral panic since the war, and that, for that reason alone, the problematic status of any evidence that is introduced into public debate is really a technicality. This particular moral panic was choreographed around the couplet 'sex and violence', but was connected, in debate within the mass media, with a much larger range of issues and behaviours (rape and sexual offences, AIDS, the censorship of videotape, gay rights, street violence, the behaviour of children in schools, drug abuse, etc.). There was certainly every good reason for the warning bells to be rung in progressive and liberal circles. There may also be some grounds for anxiety on the socialist left as such, especially the left that is seriously committed to feminism, but we do want to be more careful with that assertion.

It goes without saying that in such a heated atmosphere the reflexes of the socialist left will be to act in solidarity with those minority groups who seem threatened with some persecution, and against those who would give the State further powers to intervene into civil society, especially into the private activities of minorities. The left, after all, knows that it is one such minority group itself.

The left is also sufficiently informed to realise that this is by no means the first of such moral panics in Britain,[7] and that this is not the first time, either, in which social violence has been explained in terms of the public presence of certain kinds of visual image. The violence of the early 1950s Teddy Boys was widely attributed, by respectable commentators, to the screening of American gangster movies, while the increase in the official rates of juvenile delinquency after 1955 was thought explicable by many journalists and other social commentators in terms of the import into Britain of American horror comics.[8]

Quite justifiably, the socialist left is also suspicious of the particular pressure groups which have helped to create, and now so effectively sustain, the present moral panics over questions of sexuality and violence. Margaret Thatcher's denunciation of rape as a 'barbaric' crime, in March 1986, has in no way established the British Prime Minister as a trusted friend of the women's movement, since the movement will never forget or forgive what the Thatcher government has meant for women, across a broad range of fronts (from equality at work to day care, to the sustained destruction of the job markets for women). Mary Whitehouse's National Viewers and Listeners Association (NVALA), campaigning for further extensions of censorship (particularly of television but in relation to all kinds of other media, from video to theatre) originally emerged out of the ashes of the Moral Rearmament Movement in the early 1960s.[9] Many of the most vocal supporters of the new puritanism in Britain are also among the most virulent fractions of the radical right within the Tory Party, whose commitments extend – with Mr Norman Teb-bitt, the radical Right's spokesman formerly acting as Chair of the national party, as the most unambiguous example – to the destruction of the socialist idea and the Labour movement as such. Many socialists are surely aware, as well, of the considerable overlap between the radical right's political and economic organisations, on the one hand, and the pro-censorship, pro-family and generally morally repressive groups, on the other.

In Canada, Mr David Scott, the leader of the main pressure group arguing for censorship of prime-time television, the Campaign against Media Pornography, has his material published by a radical right pro-family group in Washington DC. The proceedings of a recent conference held in Toronto by this organisation, while claiming some openness to feminist perspectives and hearing an address by Andrea Dworkin, actually opened with prayers and a born-again address to conference participants, and proceeded eventually to listen attentively and appreciatively to Dr Everett Koop, the Surgeon-General of the Reagan administration (an administration which has done more to undermine the hospital and medicare system in the US than any other post-war government), speak on pornography and violence as pressing issues of public health.

It is not just that the socialist left is perturbed by the pro-censorship and moralistic lobbies because of the company they keep. It is also that the left is instinctively sensitive to the perverse kinds of political and social argument which groups like the NVALA have been trying to establish as credible, sensible accounts of ongoing events. In the aftermath of the dreadful events at the European Cup Final in Brussels in June 1985, for example, Mrs Whitehouse chose to develop a public account with absolutely no reference to the sources of the rabid jingoism which has been generated (for example, in the media on which the NVALA claims to be so expert) in whole sections of the working class and petit-bourgeoisie over recent years (sanctioned, officially, of course, during the Falklands/Malvinas War). She made no attempt to identify or speak to the sources of the culture of hardness which has been encouraged in the British working-class male, now so pitifully accentuated in reaction to the rapid disappearance of traditional blue-collar working-class labour markets. Nor either did she choose to speak to the present, widely-reported sense of personal and social dislocation and dismay generated by the current policies of the present government. Mrs Whitehouse suggested indeed that the proper explanation of Brussels lay *in the past*, the legacy of three decades of 'self interested liberal-humanist sentiment' that has 'beguiled' schools, universities and churches.[10]

180

The Methodology of Socialist Humanism

In addition to the anxieties expressed on the left over the directly and unambiguously repressive politics that characterise the pro-censorship and pro-family lobbies, there has always been a serious set of theoretical and methodological objections to the kinds of scholarly work that has been undertaken on the social impact of 'sex and violence'. These objections may be said to arise in two ways.

Much of the work that has been undertaken and given attention in public debates on 'sex and violence' has fallen squarely in the so-called 'effects' tradition of behaviourist psychology.[11] In this tradition, which can have both conservative and liberal variants, the key issue is whether the viewing of particular material can be shown to have particular effects on the behaviour or subsequent attitudes of the viewer. The debates of the post-war period have revolved around polar opposite interpretations of pornography and/or violent visual imagery as having a *trigger effect* or alternatively as having the effects of satiating particular individuals, perhaps in the form of a *cathartic effect*.

For liberals and progressives, there has always been some scepticism over the evidence that is adduced by conservatives in support of the trigger hypothesis. The reservations derive not so much from the results of the particular studies, which can always be unpacked in a technical way, but in the overall objection of liberals and progressives to the 'model' of human learning that is built into behaviourist psychology.

The preferred alternative model of man for liberals is a humanist one, where the individual is seen as being in some sense a choice-maker or a creative human agent (even if not always, to paraphrase another thinker of some importance on these questions, in circumstances of his (or her) own choosing). In this view, the influence of the media in producing a 'picture of the world' ought to be treated with great scepticism: no one factor could in principle have such an overwhelming function when compared to the myriad of influences from which human beings 'make choices' on a daily basis. For these reasons, one of the most widely-quoted critiques of the behaviourist studies on the effects of sex and violence argues that it is actually unhelpful as such to think of the receipt of media messages as a primarily psychological process. Media effects, this argument continues, cannot be evaluated except in terms of a much more complex social theory of an audience that is located in a particular social environment and constituted by particular socialisation processes within culture and personal biography.[12]

It has to be said that the scepticism on the left with respect to the effects of the media on the viewing audience is fairly selective. There is much more willingness on the left, especially in its work on the direct representation by the media of political and ideological 'messages', to countenance the determinative effect that such messages may have on audience perceptions. There is, for example, a very powerful literature in Britain on the influence that the racial and racist imagery of television may play in generating and sustaining the routine racism of the ordinary (white) Englishman and Englishwoman.[13] It is only in respect, it seems, of the relationship between television or videotape portrayals of sex and violence and the actual sexual practices of audiences that the progressive–liberal–humanist seems to reach for his (and, less frequently, her) humanist philosophical anthropology as a means of disavowing the deterministic protocols of behaviourist scholarship and the public–political arguments for censorship that tend to flow from them.

One way of putting this point is to follow David Holbrook, in various critical analyses he developed some fifteen to twenty years ago of pornography and permissiveness, and to attempt to identify ungrounded liberalism or permissiveness as a form of philosophical and political liberalism. Holbrook then proceeded to advance a critique of contemporary liberalism from a metaphysical position which most male socialists would find unsatisfactory expressed in and of itself (although it has to be emphasised that it would probably not create such a profound unease among many socialist-feminists). For Holbrook, the failure to discriminate against pornography marks an imprisonment in 'physicality' which prevents us from finding 'the category of life', the 'new science of being' which represents 'man's essential creativity'. The 'sexual revolution' is thus the opposite of 'liberation'.[14]

Holbrook went on during the 1970s to explore (in the company of independently-minded observers like the conservative E. J. Mishan, and the social democrat Bernard Crick) the 'false gods' that are marketed by the sex industry – revolving around a preoccupation with what is already constituted as a set of physical needs and/or 'objective' pleasures – as amongst the primary sources of a progressive dehumanisation.[15]

Holbrook's perspective has not won many left socialist adherents – undoubtedly, in part, because of the unstintingly idealist and, indeed, elitist form in which his argument was put. But his identification of the orthodox liberal–progressive defence of free choice *within existing market arrangements* (and without regard to any particular philosophical set of purposes or concerns) as 'nihilistic' is surely a powerful argument.

Other commentators have, of course, gone further, especially in trying to explore a feminist understanding of the 'effects' of pornography on the male psyche. In these commentaries, there is absolutely no question of free choice: pornography 'scripts' sexuality in a relentless process of subconscious *conditioning*, helping to sustain the misogynous and unreciprocal character of much male sexuality. For these commentators, any future transformative possibilities with respect to social, political, economic and sexual relations will involve the *systematic re-conditioning* of male consciousness.[16]

There is no way that this prospect – a veritable cultural revolution – can be thought consistently within the assumptions of market liberalism. It must surely be possible to think the question of what a socialist position might look like on visual imagery and mass entertainment, without ourselves having constantly to retreat to an identification with existing expressions of market-liberalism. It must surely be possible to see, too, that the mass availability of pornographic imagery and certain kinds of violent themes is indeed a phenomenon of *the market* at this particular, 'modernist' stage of late capitalist development. There are surely limits to the extent to which these visual products can be understood primarily as expressions of an authentic, human exploration of the discourses of 'pleasure'.

So perhaps one important reason for the left's curious silence on the pressing public question of 'sex and violence' as an intrusive visual image has been a refusal to contemplate the deterministic kinds of social explanation that appear to be bound up with the development of a critique of liberal theory. But there is surely also a real unwillingness to speak, in the manner of David Holbrook, about the *moral* and *spiritual* character of some hypothetical, future, non-capitalist society as the warrant for a critique of the existing capitalist social order and the cultural order it currently produces.

The left's relationship to such moral and spiritual terrains has traditionally been uneasy. There is also, it must be said, a much greater willingness to discuss issues like those of 'sex-and-violence' metatheoretically than there is to think prescriptively on issues of state policy, either now or in some hypothetical future. The power of recent explorations of the current wave of horror movies, for example, should not cause us to ignore the implications of these critics' metatheoretical preferences,[17] but the critical questions for socialists may lie elsewhere.

Perhaps, then, we should be concerned not so much with the analysis of the filmic texts that are produced in current circumstances of despair and/or cynicism (affecting as it does many professional filmmakers and the liberal cultures within which they work), as much as we should be trying to conceive of broad socialist alternatives not only to the present human condition, but also particular, imaginary conceptions of the organisation of the mortal life and, indeed, of practical matters of health and dying in a socialist society. It is surely an expression of the underdevelopment of a socialist moral discourse on questions of this order, that we only have available to us an image of death and ill-health as a personal terror to be experienced at the hands of a remote bureaucracy or in the rough life of the streets (in the marketplace).

The 'Video Nasties' Campaign: the Absence of any Socialist Argument

The quite understandable reluctance of the left to associate itself, in any conceivable way, with the ideologies of the new moral right or to distance itself from libertarians and progressives and the particular minority groups for whom they sometimes speak, has effectively 'silenced' the left's voice in recent political arguments as an independent, recognisable and credible perspective. So also, we have argued, has the left's failure to articulate an alternative democratic and community-based conception of what everyday life and death might look like in a socialist order. But the debates have, nonetheless, proceeded and – without a left presence – there have been specific legislative consequences, not least, in Britain in recent years, in respect of the regulation of videotape film and the marketing of sexual materials and services in general.

We indicated earlier that the expansion of VCR use in Britain is unrivalled in western countries. The development of that industry, and the very heavy reliance of the industry on videotape films with an 'adult content', was significantly helped by ambiguities in the Obscene Publications Acts. It was not until 1977 that films were brought fully within the scope of the obscenity laws. But the Criminal Law Act which did this did not specify whether the videotape was 'a record of pictures', or an 'article' under the Obscene Publications Act or television (which section 1(3) of the Act specifically excluded). This confusion forced authorities who wanted to ban such material from video stores back to the use of the notoriously difficult section 2(1) of the Obscene Publications Act.[18]

There were a number of legal actions and close press attention from 1981 onwards to the 'video nasty'.[19] In that year the Advertising Standards Authority upheld complaints against advertisements for several video films, and in 1982 there were a series of articles in the popular and serious press recognising a popular anxiety. At the time there were *no* restrictions on the age of renters in video stores and rentals were low enough for children to club together to rent videos, which they could then play on their parents' machines in their parents' absence.

In June 1982, the Director of Public Prosecutions commenced proceedings against three videos (*S.S. Experiment Camp, I Spit On Your Grave* and *Driller Killer*), but to the dismay of the National Viewers and Listeners Association did so under section 3 of the Obscene Publications Act (allowing forfeiture under a magisitrate's warrant)

184

rather than section 2 (which allows a maximum 3-year prison sentence). The cases against all three videos, along with *Death Trap*, a video that had been shown, slightly cut, in cinemas, were heard in various courts, and found guilty and forfeited. Notice was also served by the Director of Public Prosecutions that future cases would be dealt with under section 2.

There was initial resistance by the government to direct legislative intervention and an encouragement of voluntary classification by the British Videogram Association (the official organisation representing the views of the video industry), which was duly launched on 15 April 1983. However, in Prime Minister's question time on 30 June 1983, Mrs Thatcher indicated that voluntary regulation was inadequate, and that a 'ban' was required.

The Video Recordings Bill was introduced to the House of Commons in July 1983 as a private members bill moved by Graham Bright MP. It had been drafted by the Home Office. The bill passed into law in April 1984. The effect of the Act is to *outlaw* the selling or hiring of any cassette not approved by a centralised state censorship authority. It also makes it an offence to sell or rent to a child a video classified as suitable only for adults.

There has been some opposition to the redirection of government policy towards moral 'regulation', particularly from civil libertarian organisations and from the British Videogram Association. But such opposition has been articulated primarily in terms of an *unreconstructed* liberalism, built around the sovereignty of individual free choice as well as around the other increasingly discredited liberal position that posits a radical separation between the private activities of individuals and their public effects. As public debate has shifted unmistakably in the direction of restoring a sense of moral order in both public and private life (particularly in the domain of 'the family'), the appeal of liberal free market theory and also of state policies of permissiveness towards what individuals may be allowed to do, or to view, in private, has diminished.

The key references in the moral right's campaigns for regulation and censorship are to 'family life' and to the values, indeed, of a bourgeois respectability. These are powerful references, especially given the real, social anxieties provoked by the economic and social transformations currently occurring in Britain, although, as Pratt and Sparks argue, they do not in themselves undermine the considerable popular interest in questions of sexuality and hedonism.[20] And for all that the campaigns of the new moral right do help to usher in a new epoch of moral regulation and censorship, as well as to endow the state with a dangerous power over public and private media, the right's insistence on the defence of moral order retains an important and influential appeal. It has this appeal, in part, because of the absence of any public articulate socialist argument with respect to the organisation or direction of the moral life: in particular, with the conditions of a moral life that is not based exclusively around the form and the practical substance of family life in the English middle class.

Establishing New Ground

We need to retrace our steps a little. There is little doubt that there is a considerable market in Britain for the kind of violent and frequently violently sexual video-film that has come to be spoken of as a 'video nasty'. According to the British Market Research Bureau, 11% of a large sample of videotape renters, when asked to identify the theme of the last video they rented, categorised it as 'horror'. A further 12% used the closely related categorisation of 'science-fiction'. In an earlier analysis in January 1984 of videotape rentals, the BMRB had discovered that renters aged 15–19 were much more likely than others to have rented a 'horror' video on their last visit to the video store (25%). Other than horror, only comedies and classic/thrillers (at 20 and 11% respectively) even began to compete.[21] Clearly, however, some of these videos were of movies like *Franken-stein*, which should not properly be described as 'video nasties'. Probably only about thirty of the videos then available in British video stores could properly be so described.

These data neither confirm nor deny the findings of the Parliamentary Video Group inquiry with respect to video exposure of children under 15. Given the limited nature of the evidence, libertarians will tend to deny the government's claim that it is legislating on the basis of known or uncontrovertible 'facts'. There is, however, some research on the question, undertaken by the University of Lund in Sweden, which studied questionnaire information obtained from fifty 15-year-olds described as heavy users of video, which is now very common in Sweden. One of the key findings was that the 'nasty' videos were watched mainly by boys (girls seeming to concentrate on 'romantic' videos) who watched 'nasties' because of group pressure – with reportedly deleterious physical and psychological after-effects – even though they in fact preferred adventure-based videos.[22]

We do not have to accept the mechanical determinism of the 'effects paradigm' to suggest that these materials appear to play some role, and have some significance, in the reproduction of a male bravado and/or male insensibility *vis-à-vis* human suffering. We are speaking of materials that are concerned, in large part, with the eroticisation of a connection between sex and violence. Overwhelmingly, the violence is committed against women, but we do note one deference of the notorious 'video nasty', *I Spit on Your Grave*, as an attempt at a feminist offensive against rape and the male assump-

186

tions which underpin it.[23] By and large, however, the suspicion might be that the videos are of a piece with the movies released in the late 1970s, focusing on the vengeful murder by men of difficult, independently-minded or 'nagging women'. These movies (*Friday the 13th, When a Stranger Calls, Halloween, He Knows You're Alone, Dressed to Kill*, etc.) have been widely attacked by the women's movement, on the entirely plausible interpretation that the films represented a direct ideological backlash against feminism in general and strong-minded or independent women in particular.

The feminist argument against this pornography does not and should not have to depend on arguments about whether or not such material objectifies women (perhaps all human interaction involves 'objectification') or whether or not such material is likely to have effects (in the form of imitative behaviour on the part of its consumers).[24] What is absolutely crucial, especially about violent pornography, is the way it represents sexuality, women and (indeed) men in an unequal and powerful set of relationships. It is this aspect of pornography – its role in attempting to reproduce or to reinstate unequal relations of the sexes 'in the head' – which is apparently invisible to behaviourist research, and awkward for the liberal commentators who want to evaluate pornography in terms of some demonstrable social harm.[25] How could the reproduction of patriarchal relationships be seen, in patriarchal societies, as a 'social harm'? Socialist-feminism does not need to reference these liberal preoccupations as its 'grounds for intervention'.

The subjugation and targeting of women is only one element in the 'video nasties'. The violence is sometimes perpetrated against men. As we have already mentioned, *I Spit on Your Grave*, for example, graphically portrays the revenge murder committed by a woman of the four men who had violently raped her (at the beginning of the movie). The violence in *Driller Killer* is random (in this film, a young artist becomes mentally deranged when he believes he is touched by the devil and he commences to murder twelve people, both men and women, using a powerpack drill). There *is* an emphasis in many of the nasties on brutal violence against women (it is especially marked, for example, in *S.S. Experiment Camp*), but there is also a *generalised* preoccupation with the detail of extreme human pain and suffering.

What cannot be avoided is the *apocalyptic character* of the 'video nasties' (and, indeed, of many other horror movies that might not be turned down for public viewing, like *Friday the 13th*). These movies are *preoccupied* with the detailed and prolonged explanation of extreme violence and death. What else could this be for the audience other than a metaphorical exploration of *real* annihilation? How else to explain the heavy consumption of these materials in a country like

Britain, which, when it thinks about it, frequently experiences itself as a prime target in a limited nuclear war in Europe?

We have to consider that the interest in the 'video nasty' has a particular significance sociologically and historically. It is consumed overwhelmingly by young males, and as a filmic text seems to have a very different, conjunctural character to earlier horror movies in the cinema. In distinction to the film-centred tendency to subsume the video nasty into the genre of the traditional horror film, and contrary to the argument that could potentially be put extending its origins into the preoccupation with horror and death in literature, the video nasty seems to be a *new* script, characterised by nihilism and desperation rather than satire or reflexivity. In Martin Barker's words, the 'video nasties' speak with 'cynical anarchism' of 'a society without hope'.[26]

We risk a trite conclusion. If the conditions of existence of the misogynist violence of the 'video nasty' and the interest it evokes in young men are the particular social relations of a capitalist and patriarchal society, prone, as we well know, to militaristic adventures ('lock up your daughters' read the banners on the troop-ship returning from the Falklands), then a change in those social relations is a necessary prerequisite for any serious attack on the kind of violent/sexual script that surfaces in these videos. But 'futurism' is not enough. Though any attempt to legislate on 'violence' on TV and video quickly encounters a morass of definitional and political problems, and indeed the *reality* of militarism, nuclear build-up and East-West tensions which *must* be shown, this is no strategic argument for *not* accepting a broad-based democratic appraisal and possible censorship of the *imaginary* violence of video.

We are aware of the dangers in arguing for the state having any increase in its power to censor: social democrats are by no means the only constituency who would argue for increases in the power of the state in one or other area in the current political and social climate – and some of the other constituencies would not baulk at arguing for censorship of the left itself and/or its most cherished opinions. In the past, indeed, classical liberals have argued insistently that the pluralism of political and social life in 'liberal democracy' demands a minimalist level of state interference in civil society. But Arblaster shows how this liberal insistence on freedom from state interference operates only at the level of the hypothetical 'individual' arranged against a potentially intrusive or coercive state.[27]

The liberal state, historically, has frequently allowed or even encouraged the state to intervene quite fundamentally in the activities of organised labour *in general* (on behalf of capital *in general*); and its defence of individualism has frequently masked the tendency for liberals (for example, in law) to be complicit in, or to justify, the reproduction of the power of men in general over women as a whole.

At other moments, however, liberal-democratic states, and 'liberal law', *have* been persuaded to respond to demands for the protection of particular social interests and, pertinently for this argument, to the special interests of class, gender or age-group; recent changes in family law and in sexual offences legislation *cannot* be understood as exercises in classical individualism. If we take the view, as we would argue social democrats must, that the direction of state policy is an arena of struggle between social interests (and also, therefore, an arena that calls for a democratic and accountable state) – rather than being a seamless expression of the needs of Capital and Patriarchy – the call for the censorship of violent and sexist videotapes does not have to implicate social democrats in an accommodation to a *generalised* authoritarianism of the state.

References

1. See R. Shuker, '"Video Nasties": Censorship and the Politics of Popular Culture', *New Zealand Sociology*, vol. 1 no. 1, 1986.
2. D. Fisher, 'Video Cassette Recorders: National Figures', *Intermedia*, vol. 11 no. 4/5, 1983, p. 39.
3. Parliamentary Video Group Enquiry, vol. 1, 1983, paras 1, 14–15.
4. British Marketing Research Bureau, *Forte*, Quarterly Report, November 1985–January 1986.
5. Report of a Parliamentary Video Group Enquiry, Part 1, *Video Violence and Children* (London: Oasis Projects, 1983).
6. G. Barlow, A. Hill (eds.), *Video Violence and Children* (London: Hodder and Stoughton, 1985).
7. See D. Lusted, 'Feeding the Panic and Breaking the Cycle', *Screen*, vol. 24 no. 6, 1983.
8. See M. Barker, *A Haunt of Fears: The Strange History of the British Horror Comics Campaign* (London: Pluto, 1984).
9. K. Joseph, 'Britain – a Decadent New Utopia', *The Guardian*, 21 October 1974.
10. M. Whitehouse, *The Guardian*, 10 June 1985.
11. See, in particular, W. Belson, *Television Violence and the Adolescent Boy* (Farnborough: Saxon House, 1978); H. Eysenck and D. Nias, *Sex, Violence and the Media* (London: Maurice Temple Smith, 1978); V. B. Cline, R. G. Croft, S. Courrier, 'Desensitization of children to television violence', *Journal of Personality and Social Psychology*, vol. 27, 1973; J. V. P. Check, *The Effects of Violent and Nonviolent Pornography* (mimeo) (Ottawa: Department of Justice, 1985).
12. G. Murdock, R. McCron, 'The Television-and-Violence Panic', *Screen Education*, no. 30, Spring 1978.
13. See P. Cohen, C. Gardner (eds.), *It Ain't Half Racist, Mum: Fighting Racism in the Media* (London: Comedia, 1982).
14. D. Holbrook, *The Case Against Pornography* (London: Tom Stacey, 1972), p. 2.
15. See E. J. Mishan, *Making the World Safe for Pornography (And Other Essays)* (London: Alcove Press, 1973), and B. Crick, *Crime, Rape and Gin* (London: Elek, 1974).

16. A. Dworkin, *Pornography: Men Possessing Woman* (New York: Putnam, 1981); S. Griffin, *Pornography and Silence* (New York: Harper, 1981).
17. For a clear example of these preferences, see the contributions to *Body Horror, Screen*, vol. 27 no. 1, 1986.
18. See G. Parry, P. Jordan, 'Why the Booming Porn Trade Has the Law Taped', *The Guardian*, 23 February 1981.
19. The narrative presented here draws heavily upon J. Petley, 'A Nasty Story', *Screen*, vol. 25 no. 2, March–April 1984.
20. J. Pratt, R. Sparks, 'New Voices From the Ship of Fools: A Critical Commentary on the Renaissance of Permissiveness as a Political Issue', in *Contemporary Crises*, vol. 11, 1987.
21. See British Market Research Bureau, *Forte*, Quarterly Report, November 1985–January 1986.
22. C. Patmore, 'How the Boys Find Nasty Ways to Pass the Time', *The Guardian*, 26 October 1983.
23. M. Starr, 'J. Hills is Alive: a Defence of *I Spit on Your Grave*', in M. Barker (ed.), *The Video Nasties* (London: Pluto, 1984).
24. See B. Brown, 'A Feminist Interest in Pornography: Some Modest Proposals', *m/f*, nn. 5–6, 1981.
25. T. McCormack, 'Machismo in Media Research: A Critical Review of Research on Violence and Pornography', *Social Problems*, vol. 25, 1978.
26. M. Barker, '"Nasties": a problem of identification', in Barker (ed.), *The Video Nasties*.
27. A. Arblaster, *The Rise and Decline of Western Liberalism* (Oxford: Blackwell, 1984).

BARRIE GUNTER AND JACOB WAKSHLAG

Television Viewing and Perceptions of Crime Among London Residents

Introduction

In the long-standing concern about the portrayal of violence on television, stress has usually been placed on the impact of observing the behaviour of a protagonist of violence, who purportedly sets an example of social conduct for viewers (especially the young and impressionable) to follow. More recently, however, a different perspective has emerged which focuses attention more on the meanings conveyed to mass audiences about criminal and violent propensities in society by recurring patterns of victimisation portrayed on the screen. Through its repeated portrayal of violence, some researchers have argued, television cultivates distorted perceptions of the incidence of crime and violence in the real world. Such perceptions are presumed, via natural extension, to produce an assortment of emotional dispositions including fear for one's personal safety, mistrust of others, and other less specific feelings of hopelessness. Such effects are likely to be observed among those individuals who watch a great deal of television and who may therefore acquire a great deal of their knowledge about the world from it.

Empirical demonstrations of this relationship have been derived from survey data which indicate that people claiming to be heavy viewers of television exhibit different patterns of beliefs about social violence from those held by light viewers. For example, Gerbner et al. examined fear of walking in the city in their own neighbourhood at night among a sample of New Jersey school children and individuals interviewed in two national surveys in the United States. Comparing the responses of those people who claimed to watch television for four hours or more each day and those who claimed to view for fewer than two hours, Gerbner and his colleagues found that heavy viewers in all samples were consistently more fearful than were light viewers.[1]

Efforts to replicate Gerbner's findings among British samples, however, have so far largely failed. Two initial studies from the late 1970s tested relationships between levels of television viewing and personal fearfulness. Piepe et al. asked people living in the Portsmouth area to estimate the frequence of occurrence of violent incidents locally. No substantial relationship emerged between answers given and claims of viewing.[2] Wober computed a 'security scale' from responses to items concerned with perceptions of how trustworthy people are and perceived likelihood of being a victim of robbery. Results indicated no systematic tendency for heavy viewers to have lower feelings of security than do light viewers.[3]

Although the British findings have been challenged by American researchers on grounds of differences between the wording of questions and differences in the relative amounts of television viewing carried out by people in Britain and the United States, further doubt emerged about the original American results from within the United States. Re-analysis of Gerbner's data by other American researchers failed to reproduce his results and revealed problems with the original methodology.[4] Although response was made to these critiques some doubts remain.[5] Furthermore, the American/British discrepancy was reinforced in a more recent study in which Wober and Gunter found no indication of a relationship between diary measures of television viewing and fear of victimisation among respondents, in the presence of statistical controls for certain demographic and personality variables.[6]

Apart from methodological arguments, however, the theoretical position of the Gerbnerist cultivation perspective has not been universally accepted by mass communications researchers. Another view proposed by Zillmann, for example, postulates that, if anything, the effects of viewing crime drama on television should be the opposite to that indicated by the Gerbner group.[7] Because there is little reason to expect people to view material which produces aversive states such as fear, and because television crime drama invariably features the triumph of justice – the bad guys are usually caught and punished in the end – individuals who watch these programmes should find comfort and reassurance through them.

Support for this position is provided by Gunter and Wober's discovery of a positive relationship between beliefs in a just world and exposure to television crime drama programming.[8] This finding conflicts with the contention that viewing crime drama cultivates fear and mistrust and leaves open the possibility that what is cultivated instead (or in addition at least) are perceptions of a just world. It also leaves open the possibility, however, that those who believe in a just world seek to support these beliefs by more frequent exposure to crime drama on television.

Recent research has indicated that, regardless of the direction of causality inferred from correlational links between television viewing and perceptions of crime, demonstration of the nature and significance of television's contribution to these anxieties and beliefs requires a more sophisticated model of influence than that put forward by Gerbner and his co-workers. Increasing recognition has been given to three additional elements which represent important enhancements to the original cultivation effects model. Firstly, judgments about crime can have different frames of reference. Tyler has made a distinction between two kinds of judgments people make about crime.[9] First, there are judgments at the societal level which refer to general beliefs about the frequency of crime in the community at large. Then there are personal judgments which refer to beliefs about personal vulnerability to crime and one's own estimated risk of being victimised. Tyler found that these two levels of judgment were not related to each other on all aspects. He also found that societal level judgments, but not personal level judgments, were related to media experiences. Estimates of personal risk were primarily determined by direct, personal experience with crime.

Secondly, at either one of these two levels, the perceived likelihood of other- or self-involvement in crime or the concern about such involvement are situation-specific and may not be the same from one setting to the next. Tamborini, Zillmann and Bryant, for example, demonstrated that fear of crime is not an undimensional construct.[10] As well as the personal level-societal level distinction, fear of crime in urban areas was found to differ from fear of crime in rural areas. Thirdly, relationships between levels of exposure to television and perceptions of crime may be content-specific. In other words, any influence of television on beliefs about crime may depend on how much informationally-relevant programmes (that is, those with crime-related content) are watched.[11]

In the study reported here, respondents' television viewing patterns (measured in terms of proportion of viewing time devoted to different categories of programming in addition to overall amount of viewing) were related to societal level and personal level judgments about crime in a variety of locations, urban and rural, both close to home and distant from it. We wanted to find out whether societal level judgments were more closely related to television viewing than were personal level judgments; whether perceptions of crime in some settings were especially closely related to television viewing; and whether viewing of specific categories of programming, particularly those with crime-related content, predicted perceptions of crime better than did television viewing *per se.*

Method
Television viewing diaries and attached questionnaires were sent to members of a London Panel maintained at the time of this research by the Independent Broadcasting Authority's Research Department

for purpose of routine programme appreciation measurement. Diaries contained a complete list of all programmes broadcast on the four major television channels in London during one week in February 1985. Respondents assessed each programme seen on a six-point scale ranging from 'extremely interesting and/or enjoyable' to 'not at all interesting and/or enjoyable'. Endorsements thus revealed not only appreciation levels, but also how many programmes had been seen, and of which kinds.

The questionnaire consisted of two parts. In the first part, respondents were asked about their personal experiences with crime and perceived competence to deal with an attack on themselves. More specifically, respondents were asked if they personally had ever been the victim of a violent crime, and if they knew anyone who had been. They were also asked to indicate along a five-point scale ranging from 'strongly agree' to 'strongly disagree' their extent of agreement with the statement 'I could defend myself from an unarmed attacker'. The latter item was presented with 11 items taken from or based upon Rubin and Peplau's Belief in a Just World scale.[12] Some of these items were reworded in a more appropriate British idiom.

The second part of the questionnaire dealt with perceptions of the likelihood of crime and fears of personal victimisation, and was divided into three sections. In the first of these respondents were asked to estimate along a five-point scale (ranging from 'not at all likely' [1] to 'very likely' [5]) the probability that a person living in any of five locations would be assaulted in their lifetime (societal level judgments). The five locations given were London, Glasgow, Cotswolds, Los Angeles or on a farm in the United States. In the second section, estimates were requested from respondents concerning the likelihood that they might themselves fall victim to violent assault (personal level judgments) if they were to walk alone at night for a month around the area where they live, in a local park, through the streets of London's West End, through the streets of Glasgow, or through the streets of New York. They were also asked to say how likely they thought it was that they would become 'the victim of some type of violent behaviour sometime in your lifetime' and that 'you will have your home broken into during the next year'.

In the final section, respondents were asked to say how concerned they would be for their personal safety (along a five-point scale ranging from 'not at all concerned' [1] to 'very concerned' [5]) if their car broke down at night in the English countryside, if they had to walk home alone late at night from a local pub, or if they found themselves having to walk through several streets in Los Angeles at night to reach their car.

A total of 448 usable diaries and attached questionnaires were returned, giving a response rate of 47 per cent. Data were then weighted to bring the sample into line with population parameters. The demographic distribution of the sample is shown in Table I. Percentage figures in parentheses represent the known proportions for each demographic category in the London ITV region based on Broadcasters' Audience Research Board (BARB) Establishment Survey figures for 1985.

TABLE I

Demographic characteristics of the sample

| | | Sex | | Age | | | | Class | |
Total	Males	Females	16–34	35–54	55+	ABC1	C2	DE
n 448	218	230	183	139	126	211	130	106
% 100	49(48)	51(52)	41(36)	31(34)	28(28)	47(47)	29(26)	24(28)

Scoring
With regard to television viewing behaviour, each respondent was given a score for the total number of programmes watched and the numbers watched for each of nine different categories of programmes: action-adventure, soap opera, British crime-drama, American crime-drama, films, light entertainment, sports, news and documentaries/general interest.

On the basis of a frequency distribution of the total number of programmes viewed during the survey week, respondents were divided into three categories by amount of viewing: light viewers (32% of the sample), medium viewers (34%) and heavy viewers (34%). Light viewers were those who watched fewer than 25 programmes during the week, which on the assumption of an average programme duration of half-an-hour, is equivalent to less than one-and-a-half hours per day. Heavy viewers were those who watched more than 35 programmes a week (or more than three hours a day), and medium viewers were those who fell in between light and heavy viewing limits.

For each programme type, relative proportions of total viewing time devoted to each were computed by dividing the number of programmes seen in a category by the total number seen overall. This was done to obtain a more precise measure of how viewers shared out their total viewing time among different types of programmes. Frequency distributions were then computed on these viewing variables so that respondents could be divided into light, medium and heavy viewers within each programme category.

195

Results

Experience with Crime and Competence to Deal With It
Direct personal experience with violent crime was rare among this sample of London residents. Only 7% of respondents said they had ever been the victim of a violent assault themselves. Indirect contact with violent assault through knowing someone else who had been a victim was more widespread; 26% said they knew a victim.

Further details are shown in Table II, where a number of demographic differences in personal experience with violent crime can be discerned. Although men were only slightly more likely to say they had been victims themselves than were women, they were quite a lot more likely to know a victim. Age differences were apparent too. Younger people (aged under 35 years) were nearly twice as likely as older people to say they had been victims of an assault. Indeed, nearly one in ten young people said they had had this experience. Knowing a victim was equally likely across age-bands, however. Directly experienced personal victimisation was more commonplace among working-class (DE) respondents than among middle-class (ABC1) respondents.

Respondents had mixed opinions about whether they could effectively defend themselves against an unarmed attacker. Responses were equally divided between those who judged that they could defend themselves (32%), those who thought they could not (34%) and those who were unsure either way (34%). Once again, as Table II illustrates, marked differences of opinions among individuals were associated most strongly with sex and age. Men were nearly three times as likely as women to have confidence in their ability to look after themselves, while younger and middle-aged respondents had greater confidence than did older respondents.

Perceived Likelihood of Victimisation: Others
Respondents were asked to estimate the likelihood that a person living in each of five different locations would become a victim of a violent assault during their lifetime. Results indicated that greatest risk was perceived to exist for people living in urban locations. Such locations in the United States, however, held much more danger than their equivalents in Britain. As Table III shows, the place seen as potentially the most dangerous to live in by Londoners was Los Angeles. Far fewer respondents perceived a similar likelihood of a person being a victim of assault in Glasgow and central London. The locations perceived as safest of all were the rural areas, both in Britain and the United States.

Women were more likely than men to perceive victimisation as a likely occurrence for others across four out of the five locations.

TABLE II

Personal experience and competence to deal with violent assault upon oneself

		Sex			Age			Class	
	All %	Male %	Female %	16–34 %	35–54 %	55+ %	ABC1 %	C2 %	DE %
Have you ever been the victim of a violent crime?									
Yes	7	8	6	9	7	4	6	5	10
No	93	92	94	91	92	96	94	95	90
Has anyone you know ever been the victim of a violent crime									
Yes	26	31	21	24	33	25	28	22	25
No	73	69	79	76	67	75	72	78	75
I could defend myself from an unarmed attacker									
Agree	32	49	17	39	34	22	35	32	31
Disagree	34	19	47	26	29	48	32	35	35
Unsure	34	32	36	34	37	30	34	34	35

TABLE III

Perceived likelihood of victimisation for others during their lifetime*

		Sex			Age			Class	
	All %	Male %	Female %	16–34 %	35–54 %	55+ %	ABC1 %	C2 %	DE %
Likelihood of being assaulted for a person living in:									
Los Angeles	77	72	81	77	76	77	74	78	81
Glasgow	49	44	55	46	47	56	48	48	54
London (West End)	43	38	46	46	32	49	36	41	56
Farm in USA	11	7	15	13	13	8	10	9	15
Cotswolds	3	3	3	3	4	2	1	2	7

* Percentages are of those who, on a five-point risk scale, scored likelihood of assault as either 4 or 5.

There was also a marked class differential, particularly with respect to perceptions of risk in the West End of London. Working-class respondents were much more likely to perceive social danger for others.

Perceived Likelihood of Victimisation: Self
Results once again showed that perceived likelihood of victimisation varied across different locations. The scenarios painted for respondents in this section of the questionnaire once again varied along one dimension in particular – their degree of proximity to where they lived. As Table IV shows, perceived danger levels rose with increasing distance from home. By far the most dangerous place to walk alone at night, for this London sample, was New York, which was perceived to hold real risks of personal assault for more than five times as many respondents as was their own neighbourhood, where few respondents perceived any real danger.

TABLE IV

Perceived likelihood of victimisation for self*

	All %	Sex Male %	Female %	Age 16–34 %	35–54 %	55+ %	Class ABC1 %	C2 %	DE %
Likelihood of being assaulted oneself if walking after dark alone in:									
New York	83	70	87	85	81	84	84	81	87
Glasgow	53	45	59	46	55	57	49	58	52
London (West End)	41	33	53	47	38	44	33	50	57
Local Park	30	23	42	35	27	37	25	37	44
Own neighbourhood	15	10	19	12	11	20	12	13	21
Likelihood of being a victim in own lifetime	21	24	17	25	16	20	20	23	18
Likelihood of having home burgled in next year	23	23	27	20	22	36	22	26	30

* Percentages are those who, on a five-point risk scale, scored likelihood of assault or personal risk as either 4 or 5.

There were demographic differences in levels of perceived risk to personal safety. Across all locations, women more often perceived a strong likelihood of being violently assaulted than did men. The gap between the sexes was smallest with regard to perceived danger in the local neighbourhood, where it was reduced to 9%. Age was not as consistently associated with differences in perceptions of danger to self across locations. The most marked difference emerged with respect to perceptions of risk in one's own locality, where older people more often thought they were likely to become victims than did younger or middle-aged people. Class was associated with risk perceptions for self, but only with respect to more proximal locations for respondents. Thus working-class respondents were more likely than middle-class respondents to mention the possibility of danger to self from violence in central London, a local park and in their own neighbourhood. However, working-class respondents did not think of themselves as likely to fall victim to any violence in their lifetime more often than middle-class respondents.

Fear of Victimisation

How afraid were respondents of being victims of violence? To what extent did concern for personal safety vary with the location in which one might find oneself? Three items were presented dealing with fear of victimisation. Results presented in Table V indicate that respondents said they would be most concerned for their personal safety if they found themselves walking alone after dark in the streets of Los Angeles. Fear of being assaulted was mentioned twice as often for Los Angeles as in either of two British locations. Respondents

TABLE V

*Fear of victimisation**

	All %	Sex		Age			Class		
		Male %	Female %	16–34 %	35–54 %	55+ %	ABC1 %	C2 %	DE %
Fearful of walking alone after dark in Los Angeles	87	61	81	67	67	84	69	69	81
Fearful of walking alone after dark from local pub	47	30	64	41	45	60	42	44	65
Fearful of being stranded in English countryside after dark	27	13	41	24	24	35	23	26	38

* Percentages of those who, on a five-point scale of concern for personal safety, scored either 4 or 5.

199

associated the least amount of fear with being stranded after dark in the English countryside.

Demographic differences emerged associated with sex, age and class of respondents. Fear of personal victimisation was most often mentioned across all locations by women, the elderly and working-class respondents. Differences between the responses of men and women, the young and old, middle-class and working-class were quite substantial in every case.

Personal Experience with Violence and Risk Perceptions
To what extent do direct and indirect real life experiences with violence and belief in one's own ability to defend oneself against an assailant colour or mediate perceptions of social danger? As the results presented in Table VI indicate, whether or not respondents had ever been victims of violence themselves or knew someone who had been made little difference to their perceptions of the likelihood of others being victimised. Belief about one's competence to defend oneself, however, did make a difference. With respect to risk perceptions for people living in urban locations in particular, whether in Britain or the USA, respondents who felt incapable of defending themselves effectively were more likely to perceive danger.

TABLE VI

Personal experience and competence to deal with violence and perceptions of likelihood of assault for others

	Whether been a victim		Whether know a victim		Competence to defend oneself	
	Yes %	No %	Yes %	No %	High %	Low %
Likelihood of being assaulted for a person living in:						
Los Angeles	75	77	81	77	63	76
Glasgow	53	50	55	48	40	54
London (West End)	44	42	44	42	38	50
Farm in USA	8	11	12	5	11	15
Cotswolds	2	3	4	3	4	4

One might expect personal experiences with violence to have a more substantial impact on perceived environmental risks to oneself than in relation to perceptions of risk for others. The results, however, as shown in Table VII, indicate otherwise. For most scenarios neither direct nor indirect experience with violence differentiated risk perceptions relating to self. The one notable exception was in relation to the perceived chance of being assaulted in one's own neighbourhood. Respondents who had been victims of an assault before were more likely than those who had not to perceive danger near to home. Again, though, belief in one's own ability to handle trouble emerged as an important mediator of risk perceptions. Across all locations, local and distant, perceived likelihood of personal victimisation was greater among respondents who had little confidence in their ability to defend themselves.

TABLE VII

Personal experience and competence to deal with violence and perceptions of likelihood of assault for self

	Whether been a victim		Whether know a victim		Competence to defend oneself	
	Yes %	No %	Yes %	No %	High %	Low %
Likelihood of being assaulted oneself if walking after dark alone in:						
New York	85	85	88	84	73	85
Glasgow	49	53	58	51	40	61
London (West End)	49	42	38	45	36	53
Local Park	30	34	29	34	24	47
Own neighbourhood	24	14	12	16	5	26
Likelihood of being a victim in own lifetime	56	18	25	19	21	27
Likelihood of having home burgled in next year	17	26	26	24	21	33

Two more estimates of personal risk exhibited stronger associations with personal experiences with violence, however. Victims of violence were three times as likely as others to say they thought they would be victims of criminal assault during their lifetime. Clearly, and not surprisingly, the experience of victims had coloured their outlook. Indirect contact with violence, through knowing a victim, proved less powerful as a discriminator of perceptions. So too did belief in one's competence to defend oneself. Perceived likelihood of having one's home broken into was related to personal experience with violence but in the opposite direction to the above perception: victims were *less* likely to believe there was a good chance of being burgled during the next year. Perceived risk from burglary was predictably (given the above findings) greater among respondents lacking confidence in their ability to defend themselves.

Personal Experience with Violence and Fear of Victimisation

As Table VIII shows, respondents with previous experience of being victims of a violent assault were in general more concerned for their safety within each of the scenarios that had been painted for them. This factor made the most profound difference with respect to the most local of the three settings – the scenario in which respondents had to imagine themselves walking home alone late at night from a local pub. Indirect experience was a less powerful discriminator, although it did make some difference with respect to British scenarios. In contrast to direct experience, however, indirect experience with violence was associated with being less fearful. The most powerfully related variable of all was belief in one's competence in self-defence. For judgments of concern for personal safety in settings at home and abroad, respondents who felt they could not effectively defend themselves against an unarmed attacker were more concerned about their chances of being assaulted.

TABLE VIII

Personal experience and competence to deal with violence and fear of victimisation

| | Whether been a victim | | Whether know a victim | | Competence to defend oneself | |
	Yes %	No %	Yes %	No %	High %	Low %
Fearful of walking alone after dark in Los Angeles	79	71	70	72	54	79
Fearful of walking alone after dark from local pub	60	46	41	50	31	73
Fearful of being stranded in English countryside after dark	33	26	20	30	18	46

Television Viewing and Perceptions and Fear of Victimisation
The results above indicate marked variations in some perceptions of victimisation associated with certain demographic characteristics of respondents, their direct and indirect experience of assault and perceived self-defence capability. In order to find out if television viewing or viewing of specific programme types were related to risk perceptions independently of these other variables, a series of multiple regression analyses were run in which ten television viewing variables, demographics, personal experience with violence (direct and indirect), belief in ability for self-defence, and belief in a just world, were related to each risk perception. Each regression procedure was executed with all independent variables entered equally.

Table IX shows the results for perceptions of risk for others. As the table shows, in the presence of multiple statistical controls for other variables, there was only one instance of a television viewing variable exhibiting a significant relationship with a victimisation-likelihood perception.

Viewing of television news was negatively related to perceived likelihood of victimisation for someone who lives in Los Angeles. Heavier viewing of the news predicted the perception of less danger in Los Angeles for others. None of the serious drama or crime-related programme categories (for example, action-adventure, US crime-drama, UK crime-drama) was significantly related to any perceptions of risk for others. More significantly to these perceptions were whether respondents knew a victim of an assault. Respondents who knew a victim perceived greater danger for others who live in Los Angeles, Glasgow and rural USA.

Table X presents the results for similar analyses computed for perceptions of likely risk to self in different locations. Six significant relationships emerged between these perceptions and television viewing variables. Viewing of soap operas and of UK crime-drama predicted perceived risk in own neighbourhood. Heavier viewing of both programme types predicted the perception of greater danger to self in this setting. Total television viewing was significantly related to perception of potential danger in a local park and in London's West End at night. In both instances, heavier viewing predicted perception of greater risk. Finally, soap operas emerged as significant predictor of perceived personal danger if walking alone at night in the streets of New York and perceived likelihood of having one's home burgled in the next year. Heavier soap opera viewing predicted greater perceived danger in New York, but less perceived danger of being burgled.

Self-defence capability emerged most consistently as a significant predictor of perceived likelihood of self-victimisation across settings.

TABLE IX

Multiple regressions showing relationships between television viewing, personal experience with violence and demographics with perceived likelihood of victimisation for others

| | Risk for Person who Lives in: | | | | | | | | | |
| | London | | Los Angeles | | Glasgow | | Cotswolds | | Farm in USA | |
	Beta	t	Beta	t	Beta	t	Beta	t	Beta	t
Total TV viewing	.06	1.00	.02	.33	-.01	-.25	.00	.07	.02	.26
Action adventure	.11	1.43	-.07	-.87	.02	.31	-.03	.43	-.02	-.30
Soap operas	.05	.87	.10	1.69	.01	.25	.06	1.06	.03	.51
Sport	-0.44	-.77	-.04	-.83	-.05	-.91	-.09	-.173	-.01	.14
Light Entertainment	-.04	-.78	-.09	-1.53	.04	.73	-.05	-.84	-.10	-1.80
News	-.04	-.64	-.13	-2.05*	-.06	-.99	.00	.02	-.03	-.45
Documentaries	-.05	-.89	-.02	-.35	-.03	-.63	-.04	-.82	-.01	-.19
Films	-.03	-.49	-.05	.78	-.10	-1.54	.02	.30	-.01	-.20
US crime drama	-.001	.10	.14	1.93	-.00	.06	-.01	.10	.13	1.82
UK crime drama	.02	.28	-.08	-1.49	.04	.68	-.03	-.60	-.09	1.58
Sex	.09	1.68	.09	1.66	.07	1.38	-.00	-.02	.11	2.09*
Age	-.00	-.01	.00	.05	.10	1.92	-.06	-1.02	-.05	-.88
Class	.14	2.84**	.05	1.00	.07	1.28	.13	2.57	.06	1.22
Just World	-.07	-1.44	-.01	.16	.05	-1.09	-.01	.27	.02	.41
Been a Victim	-.03	-.50	-.04	-.79	-.06	-1.12	-.01	.27	.02	.41
Know a Victim	-.07	-1.25	-.16	-3.09**	-.15	-2.16**	-.01	-.20	-.12	-2.17*
Defend oneself	.07	1.40	.06	1.04	.04	.80	.09	1.61	.06	1.13
Multiple R	.30		.31		.27		.23		.27	
Multiple R²	.09		.10		.07		.05		.07	
F	2.33		2.46		1.89		1.26		1.79	
df	17/399		17/392		17/398		17/399		17/390	
p	.002		.001		.02		ns		.03	

Levels of statistical significance: *** p < 0.001, ** p < 0.01, * p < 0.05

204

TABLE X

Multiple regressions showing relationships between television viewing, personal experience with violence and demographics with perceived likelihood of victimisation for self

	Own area		Local park		London West End		New York		Glasgow		Victim lifetime		Home burgled	
	Beta	t	Beta	t	Beta	t	Beta	t	Beta	t	Beta	t	Beta	t
Total TV viewing	.01	.19	.13	2.33*	.14	2.55**	.08	1.41	.06	1.08	.06	.95	.08	1.36
Action adventure	.05	.67	.04	.51	.02	.29	.01	.14	-.01	-.09	.08	1.07	.14	1.81
Soap operas	.12	2.10*	.06	1.16	.05	.82	.15	2.70**	.08	1.38	-.01	-.25	-.13	-2.21*
Sport	-.05	-.97	-.08	-1.68	-.10	-1.92	-.02	-.45	-.07	-1.38	-.05	-.86	-.08	-1.49
Light Entertainment	-.11	-1.97	-.04	-.74	-.05	-.90	-.03	-.51	.04	.76	-.06	-.99	-.03	-.54
News	-.02	-.33	.04	.76	-.04	-.72	-.02	-.30	-.04	-.59	-.07	-1.19	-.01	-.20
Documentaries	-.07	-1.23	-.03	-.61	-.02	-.46	-.00	-.04	-.06	-1.15	-.04	-.77	-.03	-.49
Films	-.09	-1.41	-.07	-1.18	-.06	-.94	.06	.92	-.05	-.84	-.00	-.01	-.03	-.54
US Crime	-.08	-1.19	-.01	-.10	-.00	-.04	-.00	.01	-.05	-.73	-.07	-1.03	-.08	-1.08
UK Crime	.11	2.07*	.04	.83	.02	.30	-.04	-.75	.05	.96	.01	.23	.03	.62
Sex	.05	1.03	.20	3.89***	.10	1.90	.11	2.07*	.02	.31	-.07	-1.37	-.03	-.48
Age	.03	.55	-.00	.04	-.06	-1.13	-.08	-1.40	.07	1.28	-.05	-.99	.11	2.03*
Class	.10	2.08*	.13	2.69**	.22	4.37***	.04	.82	.04	.72	.00	.07	.07	1.41
Just World	-.06	-1.23	-.02	-.41	-.05	-1.03	.04	.76	-.09	-1.77	-.12	2.48**	-.07	-1.37
Been a Victim	-.02	-.41	-.04	-.82	.03	.63	-.01	-.28	.06	1.19	-.19	-3.59***	.00	.08
Know a Victim	.04	.85	.01	.20	-.02	-.32	-.17	-3.19**	-.12	-2.23*	.00	.08	.04	.68
Defend oneself	.15	2.83**	.13	2.62**	-.09	1.65	.03	.53	.13	2.50**	.03	.54	.11	1.97*
Multiple R	.33		.41		.37		.30		.30		.29		.25	
Multiple R²	.11		.17		.14		.09		.09		.09		.06	
F	2.87		4.79		3.78		2.23		2.27		2.19		1.53	
df	17/399		17/399		17/399		17/392		17/396		17/399		17/397	
p	.0001		.0001		.0001		.004		.003		.004		.08	

Levels of statistical significance: *** p < 0.001, ** p < 0.01, * p < 0.05

Greater confidence in being able to defend oneself was associated with a reduction in perceived likelihood of being assaulted.

Table XI presents the results for fear of victimisation. Heavier total television viewing was a significant predictor of level of concern in all three scenarios. Throughout, heavier television viewing predicted greater concern for personal safety. With regard to the scenario closest to home (that is, walking home alone at night from a

TABLE XI

Multiple regressions showing relationships between television viewing, personal experience with violence and demographics with fear of victimisation

	Stranded English countryside		Concern if: At night in Los Angeles		Walk home at night from pub	
	Beta	t	Beta	t	Beta	t
Total TV viewing	.11	2.09*	.13	2.23*	.13	2.51**
Action adventure	.12	1.70	.00	.09	.14	4.02**
Soap operas	.05	.97	.10	1.74	.08	1.52
Sport	−.06	−1.20	−.09	−1.65	−.15	−3.15**
Light Entertainment	−.01	.18	−.06	−1.13	−.03	−.57
News	.00	.07	−.06	−.96	.04	−.73
Documentaries	−.07	−1.51	−.07	−1.28	−.04	−.73
Films	−.03	−.54	.02	.75	−.06	−1.09
US Crime drama	−.11	−1.72	−.02	−.35	.16	2.45**
UK Crime drama	.03	.58	.01	−.15	.00	.03
Sex	.31	6.42***	.15	2.88**	.28	5.89***
Age	.03	.67	.08	1.46	.08	1.60
Class	.14	3.11**	.09	1.76	.15	2.23**
Just World	−.02	−.37	−.10	−2.02*	−.00	−.04
Been a Victim	−.05	−.97	−.06	−1.24	−.09	−2.02*
Know a Victim	.06	1.24	−.03	−.51	.06	1.34
Defend oneself	.15	3.14**	.08	1.53	.12	2.50**
Multiple R	.51		.39		.54	
Multiple R^2	.26		.15		.29	
F	8.35		3.90		9.22	
df	17/398		17/379		17/387	
p	.0001		.0001		.0001	

Levels of statistical significance: *** $p < 0.001$, ** $p < 0.01$, * $p < 0.05$

local pub) concern for safety was also predicted by amount of viewing of action-adventure, US crime-drama and sport. Heavier viewing of each of these programme categories predicted greater concern for personal safety.

Discussion

A survey among London residents investigated their perceptions of crime at home and abroad, and found that perceived likelihood of victimisation for others and for self, and concern about victimisation for self, varied with the situation, demographic characteristics of respondents, their direct experience with crime, and confidence in personal ability for self-defence in the face of an assault. Television viewing patterns were relatively weak and inconsistent indicators of judgments about crime.

Unlike the findings of Tyler, no evidence emerged here that societal level judgments (for example, perceived risks for others) were more strongly linked to media experiences than were personal level judgments about crime (for example, perceived risk for self). If anything, television viewing variables were more often and more powerfully related to perceptions of risk for self. One note of consistency with Tyler, however, was the fact that personal experience with crime was an important predictor of personal level likelihood judgments and fear of crime.

At the personal level, victimisation perceptions varied with the situation. Respondents were less likely to perceive danger close to home than in more distant situations. Furthermore, in the case of one variable, self-defence capability, its significance as a predictor seemed to depend upon the situation about which judgments were being made. Thus, lacking confidence in one's ability to defend oneself predicted the perception of greater danger to self and greater concern for safety, but only in British locations. There was no such obvious patterning to television viewing predictors of victimisation perceptions across different situations however.

Just three television viewing variables emerged from the regression analyses as significantly related to perceptions of likelihood of self-victimisation and fear of victimisation. These were total amount of television viewing, soap opera viewing and US crime-drama viewing. The latter, however, was significantly related only to one perception.

Perceived likelihood of self victimisation in one's own neighbourhood was greater among heavier than among lighter viewers of soap operas and UK crime-drama. Greater potential danger to self in a local park in London's West End at night was connected with heavier viewing of television in general. Heavier soap opera viewing meanwhile predicted greater perceived likelihood of personal attack at night in New York, but lower perceived likelihood of being burgled.

With regard to concern about being a victim of assault, however, there was some indication that television viewing was a better predictor in the context of situations closer to home. Greater fear of victimisation across all three situations was linked to heavier total television viewing. In the situation that was probably closest (geographically) to home for respondents in this survey, however, fear of victimisation when walking home alone late at night from a local pub was also predicted by heavier viewing of action-adventure, sport and US crime drama.

Researchers previously have noted the importance of content specificity in the relationship between television viewing and perceptions of crime.[13] From this observation, one would expect crime perceptions to be predicted best of all by levels of exposure to programmes with crime-related content, such as action-adventure and crime-drama. Evidence for this sort of linkage emerged sporadically in this study. Heavier UK crime-drama viewing was associated with greater perceived likelihood of self-victimisation in one's own neighbourhood. Heavier viewing of action-adventure and US crime-drama was linked with greater concern about personal safety if walking home from a local pub alone late at night.

In the context of personal-level likelihood-of-victimisation beliefs, the best programme category predictor was soap opera. Heavier viewing of soap operas predicted greater perceived danger in one's own area and in New York, but less chance of being burgled. These findings are not entirely inconsistent with the notion of content-specificity as a mediator of television's influence in social reality perceptions, however. It has been noted by several US researchers, for example, that crime had been a major theme in soap operas for a long time and that it is becoming a more prominent focus in these programmes.[14]

With respect to fear of crime, viewing of particular categories of programmes seemed to be less relevant than simply how much television is consumed overall. This may indicate that if television is the causal agent, it really does not matter which programmes individuals watch. Rather, it is general levels of exposure that are most significant. Alternatively, it could be that television is the affected agent, with viewing levels being influenced among other things by the fearfulness of individuals. Those who have greater anxieties about possible dangers to self in the social environment may be driven to spend more time indoors watching the box. Probably nearest to the truth though may be a notion of circularity in the relationship. Greater fear of potential danger in the social environment may encourage people to stay indoors, where they watch more television, and are exposed to programmes which tell them things which in turn reinforce their anxieties.

References

1. G. Gerbner et al., 'The Demonstration of Power: Violence Profile no. 10', in *Journal of Communication*, vol. 29, 1979. See also Gerbner et al., 'Television Violence Profile no. 8: The Highlights', in *Journal of Communication*, vol. 27, 1977, and 'Cultural Indicators: Violence Profile no. 9', in *Journal of Communication*, vol. 28, 1978.
2. A. Piepe, J. Crouch, M. Emerson, 'Violence and Television', in *New Society* vol. 41, 1977.
3. M. Wober, 'Televised Violence and Paranoid Perception: The View from Great Britain', in *Public Opinion Quarterly*, vol. 42, 1978.
4. See P. Hirsch, 'The "Scary" World of the Non-Viewer and Other Anomalies: A Re-Analysis of Gerbner et al.'s Findings on Cultivation Analysis, Part 1', in *Communication Research*, vol. 7, 1980, pp. 403–56, and M. Hughes, 'The Fruits of Cultivation Analysis: A Re-examination of the Effects of Television in Fear of Victimization, Alienation and Approval of Violence', in *Public Opinion Quarterly*, vol. 44, 1980.
5. See Gerbner et al., 'Final Reply to Hirsch', in *Communication Research*, vol. 8, 1981.
6. M. Wober, B. Gunter, 'Television and Personal Threat: Fact or Artifact? A British Survey', in *British Journal of Social Psychology*, vol. 21, 1982.
7. D. Zillmann, 'Anatomy of Suspense', in P. Tannenbaum (ed.), *Entertainment Functions of Television* (Hillsdale, NJ: Lawrence Erlbaum, 1980).
8. B. Gunter, M. Wober, 'Television Viewing and Public Trust', in *British Journal of Social Psychology*, vol. 22, 1983.
9. T. R. Tyler, 'The Impact of Directly and Indirectly Experienced Events: The Origin of Crime-Related Judgments and Behaviours', in *Journal of Personality and Social Psychology*, vol. 39, 1980; 'Assessing the Risk of Crime Victimisation: The Integration of Personal Victimisation and Socially Transmitted Information', in *Journal of Social Issues*, vol. 40, 1984; T. R. Tyler, F. I. Cook, 'The Mass Media and Judgments of Risk: Distinguishing Impact on Personal and Societal Level Judgments', *Journal of Personality and Social Psychology*, vol. 47, 1984.
10. R. Tamborini et al., 'Fear and Victimisation: Exposure to Television and Perceptions of Crime and Fear', in R. N. Bostrum (ed.), *Communication Yearbook* 8 (Beverly Hills CA: Sage, 1984).
11. J. Weaver, J. Walechlay, 'Perceived Vulnerability to Crime, Criminal Victimisation Experience, and Television Viewing', in *Journal of Broadcasting and Electronic Media*, vol. 30, 1986.
12. Z. Rubin, L. A. Peplau, 'Who Believes in a Just World?', in *Journal of Social Issues*, vol. 31, 1975.
13. Weaver, Wakshlag, 'Perceived Vulnerability'.
14. N. Katzman, 'Television Soap Operas: What's Been Going on Anyway?', in *Public Opinion Quarterly*, vol. 36, 1972; M. B. Cassata et al., 'In Sickness and in Health', in *Journal of Communication*, vol. 29, 1979; J. C. Sutherland, S. J. Siniawsky, 'The Treatment and Revolution of Moral Violations on Soap Operas', in *Journal of Communication*, vol. 32, 1982.

VI CRITIQUES

GUILLERMO OROZCO-GÓMEZ

Research on Cognitive Effects of Non-Educational Television: An Epistemological Discussion

Introduction

The cognitive impact of television is not a new concern of researchers in communication and education. On the contrary, it has been an issue for at least two decades. What is striking, however, is that after so many years of TV research and despite some interesting studies, the research literature provides so little knowledge about cognitive effects of *non-educational, commercial* TV. Researchers know something about what children learn that educators want them to learn from educational TV ('intended effects'), less about what they learn that educators do not want them to learn ('not-intended effects') and even less about what children learn from non-educational TV ('unintended effects'), which is what is most frequently watched. The main purpose of this paper is to explain *why* there is so little research about the *unintended*, cognitive effects of commercial TV.

Cognitive Effects of Television

Teaching involves two main aspects, intention and success, which together imply a third: consciousness. As a conscious, intentional activity, teaching has a goal: to produce at least 'a bit' of learning in the students. Independently of whether or not the desired goal is achieved, both intentionality and explicit 'trying' are present. Teaching is intended to produce conscious learning through concrete means: students are involved in a rational process. Learning, on the other hand, can take place without the learner's awareness and without the learner's participation in a teaching activity. Learning does not require intentionality. Learning can occur without the goal of learning. Teaching as a rational activity is 'restricted in manner'. In contrast, learning is not specifically restricted to the use of rational elements. Much of what students learn is acquired through the emotions. For example, in teaching, teachers can expect to achieve the intentional teaching goal, but they cannot expect to achieve *only* that goal. Yet even when teachers do not achieve the intended goal, they still produce other bits of learning in students.

Educational and Non-Educational TV Programmes

TV programmes can be classified by their intentionality. First, an educational TV programme has an intentional teaching goal and is expected to produce some conscious learning. The goal is explicit and is built on a proposition to teach something specific; it is an invitation to conscious learning. In contrast, a non-educational TV programme lacks explicit teaching goals, although other types of goals, for example, are present.

Second, in an educational TV programme, rational elements are used to bring about learning by the viewers. The objective is to allow all participants in the teaching-learning process to evaluate the proposition, to contest or accept it. Producers of educational TV programmes try to control different technical and content-related aspects of the programme in order to ensure the successful fulfilment of the learning goal.

In commercial or non-educational TV, however, this does not occur. The producer's goal is popularity not learning. In the production of advertisements there are 'learning goals' (for example, brand-name recognition), but in most instances these intentions are masked rather than made explicit to the viewer. But other learning occurs independent of the producer's intentions or the absence of them.

Like teachers, producers of educational TV programmes cannot guarantee that the desired goal is all that will be learned by the viewers; besides learning what producers want them to learn through educational TV programmes, children learn other things as well. That is, children learn not only from what is explicitly included and consistent with the teaching goal, but also from that material which is incidental. In addition, children learn a great deal from non-educational, commercial TV programmes, perhaps more than they learn from educational TV programmes.[1]

Intended and Unintended Cognitive Effects

The cognitive effects of TV are intended to the degree that they correspond with specific teaching goals. Unintended effects have been defined as those effects produced without an explicit correspondence with a teaching goal. This does not mean, however, that most of the learning outcomes in TV programming are not related to other goals (for example, information, entertainment, propaganda).

The major problem with effects resulting from non-teaching goals is that a clear relationship between these effects and other types of goals is difficult to establish. This is problematic for commercial TV effects because one cannot make judgments about the intentionality of the programmes by looking at learning outcomes. In contrast, for programmes with an explicit teaching goal it is possible to judge the relevance of this goal for educational purposes. It is also possible to evaluate learning outcomes in relation to what was taught.

Unintended effects result from both educational and non-educational TV programmes even when producers have no intentions to provoke specific learning in their audience. While some of these effects might not represent clear damage to the audience's cognitive development, not all of what children (or adults) learn from the TV screen is positive or relevant to them. In fact, some research evidence suggests that some unintended effects are clearly negative.[2]

In an extensive review of research about TV's influence on children's construction of social reality, Hawkins and Pingree report that there is empirical evidence that TV helps shape beliefs, values and attitudes. They say that TV engenders negative attitudes towards certain classes of people (doctors, the elderly, foreigners and women) as well as towards certain family values and criminal justice. They also report that TV influences sex behaviour and attitudes towards sex roles. They explain that TV contributes to the cultivation of a sexist orientation, sometimes directing children from a non-traditional perspective to a more conservative one, and sometimes vice versa. They conclude that TV '... does appear to have a significant influence on the children's construction of social reality'.[3] Besides noting the paucity of research in this field, they attribute the lack of more detailed conclusive evidence about this influence to methodological issues. For instance, they suggest that it is difficult to account for TV's influence on social reality since '... [most likely] this influence will be that of stabilising and reinforcing the status quo – something difficult to document with statistics designed to measure differences, not the absence of differences'.[4]

The extent to which the unintended effects of TV are negative is not easy to determine. In part this is because of the methodological reason suggested above. However, the major difficulty in determining whether a TV effect is negative is the value judgment implied in the term 'negative'. There are no goal statements that can be used to label a particular *unintended* effect. The researcher or social critic must insert his/her own values. In a social science that seeks to be 'value free', unintended effects pose a distinctive problem for the researcher.

The Predominant Non-Educational Character of TV

As with other media such as film and radio, the use of TV for instructional purposes does not mean that it is primarily an educational tool. Above all, TV is a general broadcasting technology. Its development was not stimulated by educational interest; mainly political and economic forces predominated.

The consolidation of TV as a profit-oriented enterprise in most modern capitalist countries has moved it further away from educational objectives. Indeed, the late Raymond Williams contends, TV and the satellite transmission industry are an essential support for current market expansion. In particular, Williams notes that there is a '... clear intention in the strongest centres to use the new communication technology to override – literally to fly over – existing national cultures and commercial boundaries'.[5] Here Williams argues that satellite TV is seen as the modern and most perfect way of penetrating national markets and cultural areas. Multinational TV corporations already use these new channels to communicate messages to cultures and countries.

Two important points can be drawn from Williams. First, the primary economic aim of TV corporations dictates what is produced and transmitted by TV. The TV corporations do include some educational programmes, though even these are subject to commercial priorities and regulations about production and time of transmission. Second, the goal of cultivating certain values (for example, consumerism) through commercial TV broadcasting is not always made explicit. While commercial propaganda is directly manifested in advertisements, the cultivation of consumerism among TV viewers is not found only in advertisements. It permeates other regular programming in subtle ways. The promotion and reinforcement of values of 'having more' as a means to happiness is but one example.

The fact that the promotion of values like consumerism is not circumscribed to programme 'A' or 'B' means that TV's effects are diffused. The consequence of diffused effects of non-educational TV programming is that any rational evaluation of learning outcomes has to go beyond single programmes. For George Gerbner and his colleagues, the major impact of TV is not the immediate one resulting from watching a concrete programme, but that impact derived from the accumulation of TV viewing. This has been referred to by Gerbner as the 'cultivation effect' of TV.[6] It is precisely the fact that there are diffuse, undesired effects for some groups that makes it even more urgent to approach them.

Unintended Educational Consequences of Viewing Commercial TV
The unintended effects of commercial TV programming would be a less serious problem if children did not watch TV. But children and adolescents spend a great deal of time in front of the TV screen: recent figures show that in the US children and adolescents not only watch several hours per day of regular, non-educational TV programmes, on the average they spend more time viewing TV than

they spend in the classroom.[7] Ronald Slaby and Gary Quarfoth suggest that

> ... a considerable amount of 'observational' learning can occur without any opportunity for children to practice what they have seen and without the occurrence of any obvious reinforcement given either to the television performers or to the children themselves.[8]

If children learn from TV, and if that learning is not explicitly related to the teaching goals of school, it is possible that TV learning contradicts or at least fails to reinforce school learning.

One major consequence of children's increasing exposure to TV appears to be that teaching activities are losing relevance for children's cognitive development and their educational process in general.[9] Another important consequence is that much of what students are being taught in schools through rational explanations and critical thinking processes may be threatened by what students learn from TV. In fact, there is some research evidence that critical thinking is counteracted by regular TV programming to the extent that values of conformism and consumerism are fostered in this type of programming.[10] Educational researchers should look at unintended cognitive effects of commercial TV because some of what students learn from TV may be what teachers do not want them to learn. Some of this learning might be an obstacle to school-based learning; other learning may include values counter to those taught in school.[11]

In sum, educational programming explicitly designed to teach generates only a small fraction of the learning outcomes from commercial TV. Research evidence suggests that unintended effects of commercial TV may have negative consequences for children's education. These result because TV is organised for profit, not learning. Relying on Dewey's theory of education, Israel Scheffler points out that if students' learning outside the teaching process represents a negative influence for this process, then it is the educator's responsibility to take that learning into account.[12]

Researchers' Concerns in the Study of Cognitive Effects of TV

A Polemical Account of the History of TV Effects Research
In its dominant form the study of TV effects originated in the United States and was then exported to other Western countries. The rise of research about TV effects in general and TV cognitive effects in particular can be understood in light of both the economic and poli-

tical determinations of the US TV industry and the development of the social and behavioural sciences.

TV became both a profitable business in itself and the means *par excellence* for advertising, which constituted one of the most important ways for market expansion in a growing capitalist system. TV obviously did not inaugurate the modern publicity era, but given the audio-visual characteristics of TV as a medium and its broad coverage, TV opened unprecedented possibilities for the conquest of potential consumers from the ever-increasing TV audience through the advertisement of old and new products and services. This situation had lasting consequences not only for the further development of the TV industry as such – technical production, programming priorities and allocation for time of transmission – but also for research priorities.

The existing techniques of marketing research and opinion surveys were increasingly used to determine consumer preferences, both quantitatively and qualitatively. As Willard Rowland notes, the need to link TV's audience with advertisements brought new challenges for researchers like Stanton and Lazarsfeld, who were already using sampling techniques of sociology for the study of press and radio audiences.[13] The demand for a better audience research tool led to the emergence of new companies to investigate audiences' opinions and preferences and to design the required tools. Funds from different sources, such as government agencies, the TV industry itself, corporations and foundations were easily available to subsidise the new research. Even researchers whose main work was at the universities combined their academic activities with research projects for specific clients within the broadcasting world.

Several consequences resulted from the association between research groups and the broadcasting industry. Perhaps the most important consequence was the development of scholarship highly determined by the needs and priorities of the broadcasting industry. This determination was not only in research methods and objectives, but most importantly in the research questions, so that – it has been suggested – questions related to both the role of TV and that of communication in society were put aside.[14] That is, researchers implicitly accepted the purposes and priorities of the owners of the broadcasting industry.

Another important, though somewhat obvious, consequence for the new scholarship in the study of effects of TV was the emphasis on methodology. The need for better research audience tools was not a matter of principles but of methods and techniques.[15] This, however, implied a particular understanding of the task of scientific research, namely, that a separation between research purposes and the definition of the research problem is possible without affecting the understanding of the object of study in question.[16] The conception of a scientific practice designed to account as much as possible for predefined objects of study found extraordinary support not only within the broadcasting industry and related fund-generating institutions, but also within the broader scientific community, by now pragmatically oriented.[17]

The development of social and behavioural sciences in a context characterised by an unprecedented technological growth and its societal changes facilitated the primary practical orientation of scientific research. A rupture took place between the traditional, theoretically-oriented European social sciences and the new American sociology.[18] The new scientific practice rejected its roots in the old tradition as it gained force and prestige under the inspiration of brilliant philosophers, sociologists and behavioural scientists.

In turning away from the European tradition, the new social sciences embraced numbers and statistics because these were seen as neutral devices, and thus useful for research demands of objectivity.[19] The most important issues for social scientists then became those stemming from the reliability and validity of data.[20] As data collection techniques became more and more sophisticated, ever-increasing emphasis was given to social science's *predictive* character. This facilitated the exclusion of certain research categories (such as social class) as well as of discussions about social *responsibility*. This exclusion was also reinforced by methodological individualism,[21] especially in behavioural psychology. This experimental methodology took the individual as the unit of analysis, thereby addressing individual differences rather than social or structural characteristics.[22]

The Pervasive Research Interest in the Study of TV Effects
Within a context highly determined by economic and technological priorities and by the practical orientation of the scientific community, the study of TV effects developed as a particular type of research. This research was based on individually-oriented experimental observation, decontextualised from historical and societal issues.[23] At the same time, it was a research whose objectives were highly dependent on technical and economic interests.[24] Consequently, the most important original interest in TV effects research was the *effectiveness* with which TV messages were being communicated to the viewers.[25] What was most important to know was if the TV audience – considered as a set of potential, individual consumers – was being stimulated or persuaded to move in the direction determined by the TV industry. That is, the major preoccupation was for the improvement of TV's ability to reach an ever-increasing audience. Within this major interest, relevant questions for TV effects

research were about what was being received by TV viewers. In line with this original concern for the efficacy of TV messages, another research question asked about viewers' use of such messages (that is, its functionality in relation to broadcasters' goals). In synthesis, the relevant questions in TV research focused on 'what' rather than 'how' and 'why' effects occur.

Over the last three decades, the effects of TV have received increasing attention. Research models, methods and techniques to account for these effects have been modified by transformations in other disciplines. For instance, researchers' interest in the effectiveness of communication in TV messages has found additional scientific support in the development of modern information sciences.

The impact of the cybernetic model on the study of effects of TV has been twofold. On the one hand, it provided a coherent rationale for focusing attention on the transmission and reception of a TV message and away from the prior determination of the message by the sender. The underlying reasoning seems to have been something like the following: given that there is a message, is it efficiently transmitted and is it effective in accomplishing the sender's objective? A discussion about the *origin* and *purposes* of the 'given' – of what is communicated through TV – was excluded.

On the other hand, the incorporation of the cybernetic model and its rationale facilitated a marriage between the emerging information sciences and learning theories. The sender-message-receiver model fitted together with the stimulus-response model of behavioural psychology. From then on research about effects of TV has been strongly associated with experimental methods. As Stuart Hall argues, the rationale behind the study of the effects was that if TV or any other medium had effects, these effects '... should show up empirically in terms of a direct influence on individuals, which would register as a switch in behaviour'.[26]

While there has been a significant advance in the way the effects of TV are investigated as well as an increasing specialisation in the type of effects studied,[27] the general original interest in the study of TV effects remains basically the same and predominates in the field. Interest in moving the TV audience in the desired direction – originally associated with advertising – permeates many other areas of research in TV effects.[28] In sum, the major interest in the study of effects of TV has been to look at the way in which the *intentions* of the communicators are accomplished, via their influence on the behaviour of individual receivers.

The Dominant Researchers' Concern about the Cognitive Effects of TV
Following the tradition predominant within the study of TV effects, the major concern among researchers of the cognitive impact of TV

has been to explore the *potential* of TV.[29] That is, attention has been given to the enhancement of TV's *intended* effects on cognitive development. Even though researchers within the mainstream approach recognise the possibility of negative effects from commercial TV as well as the fact that TV's potential is *constrained* due to the economic and legal determinants of television as a profit-oriented institution, they have focused principally on the 'positive' effects.

There are several examples which illustrate mainstream researchers' major concern for TV's positive potential. In one of the first articles which contended that television is more than 'mere entertainment', Leifer et al. proposed a way of taking better advantage of TV through the *production* of specific educational programmes enacting TV's potential.[30] Support for the production of such programmes would seem reasonable if they are viewed as an alternative in exploring TV's potential and its use for education. What is striking, however, is that from the very beginning the recommendation was seen by its own proponents as containing limitations. Other authors also have emphasised the concern for exploring the potential and positive aspects of TV and of newer video technologies.[31]

Analyses in which TV's *technical* possibilities are reviewed *in the abstract* abound in the literature.[32] How realistic is this optimism about the positive potential of TV? While it cannot be denied that the field of educational technology has benefited a great deal from the development of video technology, the issue is not whether TV and newer video technology have potential for instructional activities. Rather, the issue is that this potential has to be demonstrated in practice; that is, enacted within existing economic, political and cultural conditions.

If researchers fully controlled what is transmitted by TV, their insistence on exploring the boundaries of its potential would make sense. But this is possible only in the field of *educational* television (and related technologies). Unfortunately, educational programming constitutes a small minority effort within the vast majority of non-educational, commercial TV. The freedom to explore TV's technical potential is limited. Mainstream researchers have been concerned principally with the potential of television *vis-à-vis* learning. An alternative perspective asks about the other effects of television that are generated in the search for what TV can be made to do and how those other effects compare with intended effects.

New Proposals and the Search for the Potential of TV Revisited
Even if inadequately contemplated, the negative influence of TV has not been totally denied. Mainstream researchers seem to implicitly recognise that previous efforts to harness the potential seen in TV for educational purposes have bypassed the negative influence. For ex-

ample, it has also been suggested that parallel prosocial programming by itself is limited in its ability to counteract commercial TV's negative influence.[33] Recently, in distinction to calls for the production of more and better educational programmes, new propositions have been made which are directed to *alter* the ways in which *viewers* watch TV.

There have been basically three propositions. The first concerns the reduction of the time children spend in front of the TV screen, or the question of the TV 'diet'; the second focuses attention on the mediation of adults when children watch TV; the third emphasises the education of viewers as needing to be more critical of what is transmitted on TV.[34]

While these efforts are very interesting and could ameliorate the negative impact of TV, the responsibility for effects is inevitably displaced from the real centre – that is, the TV industry – and reassigned to viewers themselves. Independent of the displacement of responsibility, the three propositions mentioned above are also inherently limited. 'Diets' by themselves are not necessarily educational; a person reduces the amount of food, but she or he does not necessarily learn how to eat well. The mediation of adults cannot be guaranteed all the time. And the education of critical viewers has not proved to be as effective as it would seem to be theoretically. For instance, Neuman reports evidence that, although there is a difference in TV's impact on viewers according to their educational level, this difference is very small. Neuman also indicates that a person's cultural heritage seems to be more important than educational level for differentiating the effects.[35] The reason for this might be that TV influences mostly beliefs, opinions and meanings, rather than concepts or rational thinking. Again, the point here is not to deny the intrinsic value of these propositions, but to note that this value is limited in the scope of its effectiveness.

Failed Attempts to Modify Commercial TV Programming
Last, but not least, it is important to mention that there have been some attempts to go beyond propositions that seek to ameliorate the negative influence of TV. These attempts seek to change programmes, not viewers. Groups of professionals and parents have organised to study TV programming, raise criticisms and questions, and make specific recommendations to programmers and producers. One of these groups is Action for Children's Television (ACT). This group has been very active in the past few years in making public calls for reducing violent programmes – especially when children are most likely to be watching TV – and commercialism and sensationalism in news, as well as calls for increasing prosocial and educational programming for children.[36]

Groups like Action for Children's Television are important as a public forum for discussion of TV and children's education. Results, however, have been limited. So far, the major accomplishment has been to raise awareness among the general public – including children – about the negative aspects of TV and their influence on children's education. Impact on decision-making processes, priorities and characteristics of commercial TV programming, however, has been rather modest.

The critical question for the future of groups like ACT is whether they are going to be able to construct sufficient political leverage to face the TV monopolies and to modify substantially TV programming. As Willard Rowland contends, what researchers are learning from the debate about TV research is that TV is not only a social, but also a political phenomenon, and that therefore the '... only way to approach it is to engage it politically, not just to criticise it'.[37]

Whatever Happened to Policy Recommendations?
Besides groups of parents and professionals, some mainstream researchers have noted the lack of policy research in the field. The first problem with the formulation of policies for a better TV has been the kind of information generated from mainstream research. Here one should mention the case of the report about television and behaviour sponsored by the National Institute for Mental Health.[38] The Report synthesises a great deal of research evidence in different areas such as those of TV violence, cognitive and affective aspects of TV, but fails to present clear propositions for policy. Furthermore, the report assumes an explicitly ambiguous position: it recognises the relevance of the findings for designing policy and for decision-making within the TV industry on the one hand, but on the other, states the intention of not prescribing recommendations. It is up to the good will of those willing to assume responsibility to consider the findings presented in the report. The absence of policy recommendations may be ascribed to epistemological features of the report. The unit of analysis was the individual rather than the institutional level. Cook et al. suggest that the title of the report would be better if it were something like: *Psychological Consequences of TV Entertainment at the Individual Level, with Special Reference to Children.*[39]

Mainstream Research Assumptions in the Study of Cognitive Effects of TV
One explanation for the paucity of knowledge about unintended effects of TV was advanced in the previous section. In this section a different yet complementary explanation of the lack of knowledge of unintended effects of TV is developed. The focus here is on the

assumptions underlying research about cognitive effects of TV. The first objective is to make these assumptions *explicit*. The second is to examine them critically in order to show *how* it has been *epistemologically* possible to focus on the intended effects of TV while ignoring the unintended effects.

The First Assumption: 'Intentions Do Not Affect Knowing'

Within the mainstream approach, it has been assumed that the researcher's questions and purposes – whether the production of an intended or the prevention of an unintended TV effect – *do not* make an essential difference for the understanding of the effects. This assumption is related to an epistemology in which the problem of knowing an object of study is a matter of choice of methods, independent of both the particular object in question and the researcher's purpose in studying it.[40] This position implies that the decision of what to research, whether this be learning outcomes which are improvements of viewers' cognitive skills or distortions about social reality, can be made simply on pragmatic grounds, while the choice of object of study is a personal decision made by the researchers and does not affect the knowing of the object.[41]

Within this rationale, it is possible to explain all cognitive effects of TV by looking only at those effects within the researcher's scope of interest. Researchers are able to explain a TV effect by predicting its occurrence. This is to say that the successful production of a TV effect – for example, an increase in children's vocabulary through an educational TV programme – contains an explanation of why this desired effect occurred. This assumption and its related suppositions are not tenable, for at least two reasons. The first one has to do with the 'interest relativity of explanation' and the second with the 'structural divergence' of explanation and prediction.

Explanatory Relativity

According to the notion of 'explanatory relativity' (or 'interest relativity of explanation') all explanations take place relative to a 'contrast space', which is the set of alternatives pertaining to the particular question raised.[42] Potentially every question has different answers, but only one answer would meet the correct contrast space. In the case of TV, the question for mainstream researchers has centred predominantly on the production of positive effects instead of on *why* TV effects occur at all, or on effects instead of on the institution that produces them.

As discussed in the previous section, most researchers of the cognitive effects of TV have taken for granted the fact that television is a private, profit-oriented institution; they have not questioned why this is so. In explaining the effects of TV, one consequence of not

222

asking structural questions is the lack of information about 'counter-factuals', that is, what could have been otherwise.[43] This type of information is important since it reveals what needs to happen for an effect *not* to take place. The relevance of this type of information for avoiding negative unintended effects of TV is evident. For example, a structural question about how to eradicate the negative effects of TV advertisements on children could lead to challenges to the profit-oriented character of television. Instead of placing responsibility for the effect of consumerism on children, a structural explanation would place it at the institutional level, where the economic aims of the TV industry are pursued. While an explanation at the particular or 'micro' level may be correct, it is not the *whole* explanation. The decision to limit TV research in focus has resulted in a *partial* explanation, that is, the first assumption is incorrect.

Prediction and Explanation
The second reason why the first assumption is incorrect has to do with differences between explanation and prediction. As mentioned earlier, the development of a scientific practice oriented mostly towards solving practical problems gave rise to a strongly predictive social science. TV research oriented to predictions was considered very important for accomplishing the desired end of exploring TV's potential for educational and prosocial purposes. Because prediction was the goal of most experiments, it was also methodologically important. The problem is that by *predicting* the occurrence of a specific learning outcome in the viewer, researchers have not necessarily *explained* why that learning outcome occurred.

That a successful prediction does not imply an explanation of why an event occurs is clear in the case of TV. For example, the production of a particular educational programme requires several different elements: the technology of transmission; a content to transmit; a symbol system; didactic factors and organisational factors. All of these elements somehow contribute to produce a specific effect. What a successful prediction would imply is that the necessary factors for a learning outcome to occur were present and/or rightly combined in the particular programme. As Gavriel Salomon and Alicia Martin del Campo have noted, the success of an educational programme such as *Sesame Street* in producing the desired learning outcomes in the viewers *might be* a result of '... a combination of well functioning technology of transmission and a symbol system'.[44] However, it also could be argued that success was due to other factors or other combinations. The point here is that to provide grounds for why an effect *must have* occurred is not to explain it.

The focus on predictions within mainstream TV research is only understandable in terms of the major concern among researchers of producing positive effects, for which prediction rather than explanation would be required primarily. This is because predictions have been functional to the predominant type of scientific practice. But predictions are limited even within this practice. First, predictions need to be based only on those elements which are always present when the effects occur. But what about non-events 'caused' by the absence of certain elements?

The second reason why predictions might be limited for the understanding of TV's cognitive effects is due to the types of elements which can be subject to control. For instance, in the case of educational programmes with their teaching goals, the occurrence of expected learning outcomes would depend on success in adequately controlling variables such as TV's symbol system, format of the programme, language, etc. However, in the situation when not even specific goals are known and all elements tend to escape researchers' control – which is the most common situation with commercial TV – the role for prediction is minimal.

One consequence of the emphasis on predictions rather than on explanations is the fact that educators and researchers alike can be successful in the production of desired TV effects only if certain conditions are fulfilled. The success, however, does not necessarily imply an understanding of the crucial cause of these effects. Another and perhaps more important consequence of the focus on predictions is that attention has been paid to controllable effects. Variables of those few sought-after effects of *educational* TV (for example, vocabulary expansion) are subject to control by the technical staff, unlike unintended effects of these programmes. Moreover, since decision-making about content, themes and issues to be broadcast in commercial TV is not a target of inquiry in the TV effects field, effects are left unexplored and unexplained, and not understood.

In the light of the above discussion, if *understanding* is the objective, it would make more sense to look for *explanations* rather than to focus solely on predictions. Explanations would provide information which helps to *avoid* the unintended and negative effects. In addition, a description of the *process* by which the effect comes to pass would be necessary. This would allow the researcher to know the *mechanisms* by which a set of different aspects or contingencies are related and thus cause the effect in question.[45] This discussion, however, leads to the debate between 'positivist' and 'realist' accounts of causality, a topic which exceeds the scope of this paper.

Second Assumption: 'One Can Understand TV Without Looking at Its Social Origins'
The mainstream approach to cognitive effects of TV has also assumed that it is possible to fully understand these effects by look-

ing at TV only as a medium. Some authors have emphasised that the effects of TV should be understood in the strictly technical aspects such as visual and auditory features like sound and camera shots, while others have stressed the organisational aspects and formats of programmes.[46] But underlying all these variations there is an assumption that the source of the effects is inherent in TV as a technical medium, and that therefore, effects can be understood correctly without looking at the institutional aspect of TV. This way of understanding the effects of TV has its origins in the McLuhan tradition for which 'the medium is the message'.

The belief that it is possible to understand TV and its cognitive effects without looking at the way TV as an institution is determined and controlled in society has been sustained by what Williams calls 'technological determinism'.[47] Here it is implicitly assumed that the way TV has developed and is used in society is the only possible way. That is to say, that TV developed the way it did by historical necessity and not as a result of cultural, political or economic forces representing the actions of intentional human beings.

Alternative technological developments based on other types of ownership and message distribution are indeed feasible.[48] What is important to stress here, however, is that interesting alternatives to media operation in society do already exist. Yet these alternatives are not systematically nor seriously considered in the mainstream exploration of TV's potential. Because it was taken for granted by most mainstream researchers that television and newer video technologies developed 'naturally', the medium's *raison d'être*, its specific uses and control never became problematic and thus were never legitimate objects of intellectual injury.

The Cultural Determination of TV Content and Formal Features
It was argued above that the separation between TV as institution and TV as medium is inappropriate and impossible since TV as an outcome of technical development is also the result of a political-technological process within a particular social milieu. TV cannot be conceived of only as a medium. The second reason why this is an arbitrary separation is the *inescapable* cultural determination of TV content and formal features.

That TV content is culturally influenced in a particular manner is perhaps not so difficult to comprehend. In a similar way, TV's technical aspects encompass a cultural influence. This is mainly so because culture fixes sets of rules and patterns or 'schemas' which determine a specific organisation of images and concepts, leading to the manifestation of 'codes', in TV programming. The process of encoding is a process by which TV's producers make different images intelligible.

225

Along the same line, 'professionalism' within media in general and TV in particular cannot be taken as the pure mastering of all the skills necessary to technically produce a message. Professionals working in TV industries – both in the production and in the transmission of TV messages (reporters, commentators, camera-people, etc.) – are carrying out not only a technical professional (thus neutral) performance, but are sources of cultural connotations for the TV programmes as well. This might not be by conscious intention so much as it is because all persons are immersed in a particular cultural milieu.

Third Assumption: 'TV's Cognitive Effects are Predominantly on Skills'

The predominant belief that the medium is the message led researchers within the mainstream approach to think that a separation between TV's institutional and technical aspects could be made without undermining the understanding of the effects. This led people to the further assumption that most cognitive effects of TV are effects on skills and knowledge rather than on beliefs. In other words, not only has TV been reduced to its technical aspects, but its cognitive effects have been reduced to effects only on certain cognitive domains.

This assumption has likewise been associated with a conception of education as mainly the development of skills. With TV understood as a set of technical properties, it is a simple step for the technical capabilities of each medium to be related with specific cognitive effects. Most of these effects are seen as *positive* ones in that they generally represent improvements of the viewer's cognitive skills.

The third assumption cannot be tenable for two major reasons. One, the conception of education mainly as a process of skill development is a narrow way of understanding both the learning process and education. Two, the chief characteristic of TV of presenting information *as if* it were true influences viewers' *beliefs* perhaps more than any other cognitive domain (that is, knowledge and skills).

The Cognitive Domains Affected by TV

But education is not limited necessarily to the development of skills. It certainly encompasses skills development, but it is also a *socialisation* process within certain cultural values and beliefs. Cognitive development, besides including cognitive skills such as induction and deduction (necessary to process information), also encompasses the development of conceptual knowledge, and most importantly, the acquisition of beliefs.

Israel Scheffler's distinction between skills, knowledge and beliefs is enlightening here.[49] According to him, each implies a different type of cognitive process: skills are 'knowing how', distinct from 'knowing that' and from 'believing that'. To believe something implies *to consider* what is believed to be true (independently of whether or not what is believed *is* true). This is not the case for 'knowing how' or 'knowing that'. The truth condition of knowledge does not come from the skills for arriving at that knowledge since truth is not inherent in a procedure. This means that in the case of TV viewing, truth is not to be found in the mechanical processing of TV's information, but instead in what that information proposes. For the purpose of this paper then, cognition skills constitute only a part of the cognitive domain affected by television.

The Power of TV to Make its Messages Believable
Among visual media, TV is perhaps the most powerful medium because it presents information *as if* it were true. This power derives from a number of sources. One is the service TV performs in making viewers indirect witnesses of events for which they do not have first-hand knowledge or experience. The content of TV is especially vulnerable to be taken for granted because the visual system of signs upon which this content is constructed is widely available in any culture. The availability of visual signs appears to involve 'no intervention of coding, selection or arrangement'. Furthermore, TV appears to reproduce only the 'actual trace' of reality in the transmitted images. Here one can argue, as Hall does, that this apparent reproduction is an illusion – the 'naturalistic illusion' – because the construction of TV content, that is, of the '... combination of verbal and visual images which produce the effects of reality, requires the most skilful and elaborate procedures of coding'.[50]

The coding is noticeable for those events which somehow appear to threaten the status quo, such as events related to minorities or to other countries. But for other events (for example, the roles in which women are presented in entertainment programming), the codes cannot be apprehended easily since they are inherent in a generalised way of making sense of reality. This is why Stuart Hall refers to the 'natural' as only a 'naturalised' TV re-presentation of reality.[51] That is, TV makes events' meanings reflect the most universal common sense. This appears to be a very spontaneous function of TV owing to the documentary character that the visual image possesses.

A related source from which TV acquires power to make its content believable is its high degree of 'representationalism', that is, the technical capability to represent reality. The degree of representationalism clearly increases when the technical power to *duplicate* and *replicate* increases.[52] Since the sense of 'verisimilitude', that is, the appearance of truth, is interrelated with the degree of representationalism, the higher this degree, the more verisimilitudinous the

TV content will be. The appearance of truth is, however, not only dependent on the technology of the medium in question, but also on the programme creator's intentions. This is to say that a high degree of representationalism is necessary for the appearance of truth, although it is not sufficient. This is also why different formats (for example, comedy and drama) used in TV vary in degrees of representationalism. In short, the representational character of TV is technology-bound, but the decisions about what to represent and how to do it are not. For all these reasons, the assumption that TV's effects are almost exclusively on skills is not tenable.

In sum, the assumptions examined above have, on the one hand, led to a double reductionism in the study of TV cognitive effects. On the other hand, these assumptions have facilitated researchers to *exonerate* television and the newer media from responsibility of unintended and not-intended effects on children's cognitive development, especially for those effects on beliefs. The original separation of researchers' intentions from the research object of inquiry has been reproduced at different levels. As shown above, these divisions and separations are: one, between TV as social institution and TV as technical instrument; two, between TV's cultural and technical forms; and three, between effects on cognitive skills (and knowledge) and effects on beliefs. Erroneous assumptions have meant that unintended and not-intended effects have been attributed to the 'social environment', and to TV viewers' 'malpractice' while television has been allowed to 'stay clean'.

References

1. For a pioneering study, see A. Martin Del Campo, A. Montoya and M. A. Rebeil, *El impacto educativo de la television commercial en los estudiantes del Sistema Nacional de Telescundaria, Mexico* (Mexico: Consejo Nacional Tecnico de la Educacion, 1982).
2. See R. C. Hornik, 'Television Access and the Slowing of Cognitive Growth', in *American Educational Research Journal*, vol. 15 no. 1, Winter 1978, and Gavriel Salomon, 'Television and Reading: The Role of Mental Effort Investment', paper presented at the National Policy Conference on Children and TV, Simon-Fraser University, Vancouver, Canada, 1982.
3. R. P. Hawkins and S. Pingree, 'Television's Influence on Social Reality', in National Institute of Mental Health, *Television and Behaviour: Ten Years of Scientific Progress and Implications for the Eighties, vol. 2, Technical Reviews* (Rockville, Maryland: NIMH, 1982).
4. Ibid.
5. R. Williams, *The Year 2000: A Radical Look at the Future and What We Can Do To Change It* (New York: Pantheon Books, 1983).
6. G. Gerbner, 'Television in American Society: Introductory Comments', in National Institute of Mental Health, *Television and Behaviour*.
7. See R. M. Liebart, J. N. Sprafkin and E. S. Davidson, *The Early Window: Effects of Television on Children and Youth* (New York: Pergamon Press, 1982).

8. R. Slaby and G. B. Quarfoth, 'Effects of Television on the Developing Child', in B. W. Camp (ed.), *Advance in Behavioural Paediatrics*, vol. 1, 1980, p. 237.
9. See Lloyd N. Morrissett, 'Television, America's Neglected Teacher', in *Television and Children*, Fall 1984.
10. See Montoya and Rebeil, *El impacto educativo*.
11. See Josephine Holz, 'The "First Curriculum": Television's Challenge to Education', in *Journal of Communication*, vol. 31 no. 1, 1981.
12. See I. Scheffler, *Four Pragmatists: A Critical Introduction to Pierce, James, Mead and Dewey* (New York: Humanities Press, 1974).
13. See W. D. Rowland, *The Politics of TV Violence* (Beverly Hills: Sage Publications, 1983).
14. See Montoya and Rebeil, *El impacto educativo*.
15. See W. D. Rowland and B. Watkins, *Interpreting Television: Current Research Perspectives* (Beverly Hills: Sage Publications, 1984).
16. See A. Bryman, 'The Debate About Quantitative and Qualitative Research: A Question of Method or Epistemology?', in *British Journal of Sociology*, vol. 26 no. 1, 1954; and Alan Garfinkel, *Forms of Explanation: Rethinking the Questions in Social Theory* (New Haven: Yale University Press, 1981).
17. See Huber W. Ellingsworth, 'What's Ahead For Programs in Communication? A Public Conversation', paper presented to Dallas Conference of the International Communication Association meeting, 1983.
18. See Karl Mannheim, *Ideology and Utopia* (New York: Harcourt/HBJ, 1936).
19. See Rowland, *Politics of TV Violence*.
20. See P. Gonzalez Casanova, *The Fallacy of Social Science Research* (New York: Pergamon Press, 1981).
21. See Paul Ricoeur, *Main Trends in Philosophy* (New York: Holmes and Meier, 1979).
22. See Garfinkel, *Forms of Explanation*.
23. See Rowland and Watkins, *Interpreting Television*.
24. See Thomas Cook, Deborah A. Kendzierski and Stephen V. Thomas, 'The Implicit Assumptions of Television Research: An Analysis of the 1982 NIMH Report on Television and Behaviour', in *Public Opinion Quarterly*, vol. 47, 1983.
25. See Montoya and Rebeil, *El impacto educativo* [...].
26. Hall, 'The Rediscovery of "Ideology": Return of the Repressed in Media Studies', in Michael Gurevitch et al. (eds.), *Culture, Society and the Media* (New York: Methuen, 1982).
27. See, for example, for a study on effects on attention and comprehension, Jennings Bryant and Daniel R. Anderson (eds.), *Children's Understanding of Television* (New York: Academic Press, 1983).
28. See Everett Rogers, *Communicacion en las campañas de planificacion familiar* (Mexico: Editorial Pax, 1976).
29. See Aimée Dorr, 'Children's Reports of What They Learn From Daily Viewing', paper presented to Biennial Meeting of the Society for Research in Child Development, San Franciso, March 1979.
30. Leifer et al., 'Children's Television: More Than Mere Entertainment', *Howard Educational Review*, vol. 44 no. 2, 1974.
31. See Patricia M. Greenfield, *Mind and Media: Effects of Television, Computers and Video Games* (Cambridge, Mass.: Harvard University Press, 1984).
32. See Shashi Rao, 'Communications Research in the Third World: A Review of Criticisms and Content Analysis of an Asian Research Journal', paper presented at the International Communication Association Annual Conference, Hawaii, 1985.
33. See J. Johnston and J. Ettema, *Positive Images: Breaking Stereotypes With Children's Television* (Beverly Hills: Sage Publications, 1982).

34. See, respectively: Slaby and Quarfoth, 'Effects of Television'; J. L. Singer and D. G. Singer, 'Learning How To Be Intelligent Consumers of Television', in Michael J. A. Howe (ed.), *Learning From Television: Psychological and Educational Research* (New York: Academic Press, 1983); Bryant and Anderson, *Children's Understanding of Television*.
35. W. Russell Neuman, 'Television and American Culture: The Mass Medium and the Pluralist Audience', in *Public Opinion Quarterly*, vol. 46, 1982.
36. See Action for Children's Television, *Fighting TV Stereotypes: An ACT Handbook* (Massachusetts, 1983).
37. W. D. Rowland, 'Continuing Dilemmas in Media Reform and Democratization: The Special Problem of Liberal Visions For Changing Television', paper presented at International Communication Association Annual Conference, Hawaii, 1985.
38. National Institute of Mental Health, *Television and Behaviour*.
39. Cook et al., 'Implicit Assumptions of Television Research'.
40. Jennifer D. Slack and Martin Allor, 'The Political and Epistemological Constituents of Critical Communication Research', *Journal of Communication*, vol. 33 no. 3, 1983.
41. Bryman, 'Debate about Quantitative and Qualitative Research'.
42. See, respectively: Garfinkel, *Forms of Explanation*; and Hilary Putnam, *Meaning and the Moral Sciences* (Boston: Routledge & Kegan Paul, 1979), and *Reason, Truth and History* (Cambridge: Cambridge University Press, 1981).
43. Garfinkel, *Forms of Explanation*.
44. G. Salomon and A. Martin del Campo, 'Evaluating Educational Television', in Wayne Holtzman and I. Reyes Lagunes (eds.), *Impact of Educational TV on Young Children*, UNESCO Educational Studies and Documents, no. 40.
45. See Russell Keat and John Urry, *Social Theory as Science* (London: Routledge & Kegan Paul, 1982).
46. See A. C. Huston and J. C. Wright, 'Children's Processing of Television: The Informative Functions of Formal Features', in Bryant and Anderson, *Children's Understanding of Television*; M. L. Rice, Aletta C. Huston and John C. Wright, 'The Forms of Television: Effects on Children's Attention, Comprehension, and Social Behaviour', in M. Myer (ed.), *Children and the Features of Television: Approaches and Findings of Experimental and Formative Research* (New York: K. G. Saur, 1983); A. Collins, 'Interpretation and Interference in Children's Television Viewing', in Bryant and Anderson.
47. Williams, *The Year 2000*.
48. See, for example, Williams, *The Year 2000*; and Armand Mattelart and Jean-Marie Piemme, *La Télévision Alternativa* (Spain: Editorial Anagram, 1981).
49. See I. Scheffler, *Conditions of Knowledge: an Introduction to Epistemology and Education* (Chicago: Chicago University Press, 1983).
50. Hall, 'Rediscovery of "Ideology"', p. 76.
51. S. Hall, 'Recent Developments in Theories of Language and Ideologies: A Critical Note', in Hall et al., *Culture, Media, Language* (London: Hutchinson, 1980).

MARTIN JORDIN AND ROSALIND BRUNT

Constituting the Television Audience: Problem of Method

Charlotte Brunsdon and David Morley's *Nationwide* study has become something of a classic or model text in the field of television audience research.[1] In particular, Morley's *The 'Nationwide' Audience* holds out the possibility of escaping both the exaggerated textual determination of some Althusserian or 'effects' studies and the insufficiently sociological accounts of audience response available from 'uses and gratifications' or certain structuralist approaches. Morley's work attempts systematically to refer the various ways in which different audience groups make sense of a television programme to the social relations defining their material conditions of existence – without reducing the former to a mechanical reflection of the latter. Morley has subsequently described the objective of this research as the construction of an 'ethnography of reading' and it indeed possesses wide-ranging and fundamental connections with the notable ethnographic tradition of British cultural studies. This ethnography is concerned with investigating the independent cultural systems which people themselves actively create out of, and use to mediate between, received cultural and ideological materials and the material relations within which they exist.

In 1984, we ran an ESRC-funded research project on the media's role in shaping public opinion during the Chesterfield by-election. As one part of this wider research, we attempted an informal replication of the *Nationwide* study. This involved the screening and discussion of a *Newsnight* election broadcast with around a dozen audience groups drawn from the Chesterfield community. While this research did not correspond precisely either to Morley's study nor to the alternative approach suggested here, the arguments of this study are drawn largely from our experience at this time.

The attempt to broadly follow the *Nationwide* study in our own audience research required us to pay particular attention to the way this theoretical framework had actually been realised in the design, execution, interpretation and presentation of the empirical research itself. Our aim here is to explore what we now perceive as, if not

231

quite a contradiction between theory and method in Morley's work, then at least the uneasy coexistence of two significantly different approaches to audience study. To simplify, we would say that a largely qualitative, ethnographic theory is concretised and applied through a largely formalist methodology derived from mainstream quantitative research in the social sciences. This tension threatens to obscure and undermine the value and distinctiveness of the work. We suspect that this is by no means an uncommon phenomenon in qualitative media studies. If we have concentrated on Morley's work, then, it is because of its importance as a model or point of reference and with the aim of clarifying and confirming that importance rather than wishing to discredit it. Finally, we have found it useful to approach the discussion by focusing on the use, nature and status of the decoding group, as representing a key interface between theory and methodology.

Why Not Work With Individual Respondents?

Morley's explicit answer to this question – 'much individually based interview research is flawed by a focus on individuals as social atoms divorced from their social context'[2] – is, interestingly, a negative and somewhat ambiguous justification for using groups. He discusses several representatives of such research, but the classic and most extreme example is probably one he hardly mentions – the public opinion survey as employed in the audience research of the broadcasting authorities themselves. It is not difficult to see why such methods seem to have little to interest ethnographic analyses aimed at uncovering the real processes by which audiences' decodings of television are linked to the social relations they inhabit. Shorn of its considerable sophistication and complexity, the public opinion approach is essentially formalist, measuring its adequacy by the extent to which responses are successfully decontextualised.

The method is typically addressed to formally free and equal individuals, the freedom of action/response of the isolated individuals being the guarantee of their formal equivalence and interchangeability. Respondents are thus interviewed individually to avoid the illegitimate interference of group or other contingent pressures with the freedom of response. Interviews are narrowly restricted to a set of questions uniformly applied to each respondent and carefully designed to avoid ambiguities of meaning in either question or response. Answers are thus rendered formally equivalent (and can be coded and quantitatively processed) since they have been protected from the interference of contingent or contextual differences affecting the individual researcher, interview or respondent. These answers can then be aggregated to describe collective responses, provided that the selection of respondents is organised to represent proportionately the macroscopic existence of whatever have been identified as the pertinent variables (the representative sample). These variations between otherwise free and equivalent individuals can thus be formally correlated with variations in their otherwise free and equivalent responses. Responses may either be aggregated to provide a simple mass response embracing all variables or comparatively aggregated in terms of any one or more variable(s).

The method itself is not concerned with the initial discrimination between which variables are perceived as pertinent and which contingent, this being the responsibility of a theory viewed as wholly external to the method. It is a matter of indifference to the method whether audience responses are correlated with gender, age or shoe-size. Nor is it concerned with the characterisation of the pertinent variables – how to define differences in social class or ethnicity, etc. Finally, it is not concerned with describing, yet alone explaining, causal relationships between response and variable – merely with formal (statistical) correlations which may or may not represent causal or relevant connections. It does not therefore offer an account of determining relationships in the real world (though it may be used for this purpose), but a formal 'model' of the real, made out of statistical relations between data. The adequacy of the data and the model to the real is dependent on the adequacy of the theoretical brief given from elsewhere.

In most audience research of this kind, this theoretical brief is rarely made explicit. Commonsense perceptions derived from and reproducing consensual, liberal-pluralist ideologies are simply built into the methodology as natural, unquestioned assumptions or, in a self-referencing circle, as necessary conditions of methodological rigour – that society can indeed be legitimately regarded as an aggregate of formally free and equal individuals; that their behaviour can best be understood as a function of technical, demographic variables of class, age, gender, ethnicity, taken simply as empirically observable differences between individuals; that individuals share a common, consensual culture which renders meaning unproblematic and transparent. There would thus seem to be little or no common ground between such approaches and the theoretical framework and objectives of an audience research like Morley's based on the encoding/decoding model developed within cultural studies.

Why then have we bothered to spell out a kind of research which is so widely used and familiar within the human sciences and which appears sufficiently crude in its empiricism as not even to merit serious discussion in Morley's account of the traditional paradigms of audience research? It is essentially out of a feeling that qualitative media research is often prone to reject this approach out of hand, unthinkingly.

As we have learnt from Althusser, the simple dismissal of inadequate positions without a thorough settling of accounts can often result in those positions continuing to function unperceived in the work which is most critical of them. We have said that this seems to be the case with *The 'Nationwide' Audience*, as the terminological continuity in Morley's references to 'demographic variables', 'sample', etc., might suggest. Firstly, the rejection of theoretical individualism and the absence of statistics is no guarantee of replacing the formalism of a survey approach with a materialist analysis. Secondly, it is not immediately obvious in what sense Morley's actual research does in fact analyse how *groups* decode television; and whether it could not equally well have been accomplished by working with individual respondents.

Is Group Work a Necessary Condition of Applying the Encoding/ Decoding Model?

The necessity for studies of audience decodings to use group respondents has already been questioned. Noting Morley's criticism of the way previous research has often decontextualised the individual, Justin Wren-Lewis argues that Morley is 'quite wrong to imply that an "individually placed interview" necessitates the exclusion of "social context" while a group-based interview does not'.[3]

We, too, would affirm that the individual is not a 'social atom', the simplest irreducible element of social life but, on the contrary, the most complex element, the point at which a multitude of shifting social and cultural determinations converge; and at a level of particularity in their concrete combination that defines the unique 'biography' of each individual. In principle, then, since the characteristics of individuals are not opposed to 'society' but manifest their social existence, there would appear to be at least no absolute objection here to the use of individual interviews.

Indeed, for Wren-Lewis their use is obligatory: 'It is only by interviewing each subject (as a product of a whole range of societal variables) individually that analysis can establish societal variables within decoding practice.' While his defence of this assertion touches on a number of crucial issues, however, they are somewhat misrepresented by being raised within an empiricist framework.

His argument starts from a commonsense naturalism: 'since the first stage of decoding television usually takes place on an individual basis or familial basis, it is appropriate to interview audiences on that basis'. This is neither here nor there: 'appropriate' is a long way from 'necessary', while families are hardly individuals. Having already selected the individual as the respondent, however, Wren-

Lewis can now reject group interviews for 'interfering' with the individual response – in line with public-opinion orthodoxy:

> If the members of a decoding group move towards a unifiable position, it is difficult to gauge how much this is an effect of group dynamics. Do, for example, certain dominant members of the group structure the readings of the group as a whole? The individually based interview circumvents this problem.

Despite the fact that group 'interference' becomes a meaningless notion if decoding is a group activity, there are real issues concealed here. The first is in the reference to the unifiability of a group's decoding. The second is in the earlier reference to 'the first stage of decoding television' and the admission that 'it would be even more revealing to trace the decoding subject's path through "the condition in which actual opinions are formed, held and modified"'.[4] This at least raises the possibility that decoding might sensibly be considered a more extended process than the individual's initial reading of a text and a process only possible through social interaction. We shall return to both these points later.

More fundamentally, much of Wren-Lewis's article involves taking Morley to task for pre-determining the results of his research. This, it is argued, follows firstly from Brunsdon and Morley having started with analysis of the text's 'preferred reading' – 'the sense in which a text can be seen to be organised in such a way as to narrow down the range of potential meanings that it can generate'.[5] For Wren-Lewis this preferred reading should, in a naturalistic way, be the reading preferred by the readers, constructed *a posteriori* from the range of individual decodings: 'the analyst's task is to construct a series of "preferred readings" from the material gathered *after* the interviewing has taken place'.[6]

Secondly, 'the data itself is weighted by the organisation of groups in terms of socio-economic variables'.[7] Morley himself acknowledges having concentrated in practice on class to the relative exclusion of other determinants (as he also admits subsequent doubts on the nature of preferred readings). But Wren-Lewis's criticism is broader. For him, to work with groups is to have to select or organise them and the criteria used in this effectively pre-judge what are the pertinent variables shaping decoding. The solution is to use individual interviews and avoid the problem, drawing up a list of relevant variables after the event to explain the variety of their decodings.

Again, genuine issues are here poorly formulated. Working with individuals does not avoid the need to pre-structure respondents if a representative sample is required for the results to have general validity. More importantly, both of Wren-Lewis's criticisms are manifestations of the empiricist illusion that theory must only intervene after the event to explain 'facts' otherwise untouched by human hand. Morley's privileging of certain variables (class or gender over eye colour or height) is not, as Wren-Lewis has it, a mistake attributable to human fallibility but, as Morley has it, the application of a theoretical framework derived from accumulated empirical evidence. Theory does not meddle with the authenticity of the facts; it is what enables data to be unearthed and recognised.

What is really at stake here is not the empiricist opposition between theory and facts but the relationship between different levels of analysis and abstraction, and how one moves from the empirical to the general. While the choice between the individual or the group as the focus of study is a real and important one, it involves the more fundamental question of whether the chosen 'object of study', is analysed in its own right as a specific complex *structure* of multiple determinations, or as formally representative of the operation of one or more conceptually separable, social structural factors.

What Do Groups Represent?
By considering the problem of how to generalise from empirical data, we can thus further clarify the still open question of whether the kind of audience research proposed by Morley requires or 'prefers' individual or group respondents. In the case of individual interviews, for example, how is one to interpret an individual's reading of a television text, given that she/he is to be seen as the product of multiple social determinations? How is one to generalise the reading made by, say, a black, working-class woman so that one can discriminate between the respective effectivities of ethnicity, class or gender?

Set up this way, the problem calls for a comparative study, firstly of the readings made by other black, working-class women; secondly, between these and the readings made by black, working-class men (the differences representing the effectivity of gender on decoding); then between both these and the decodings by white, working-class women and men (revealing 'ethnicity') and so on. That is to say, the procedure involves the two moments of grouping together formally equivalent individuals to define the commonality in their decoding; and of comparing two groupings, differing in respect of one variable, to define the differences in their decodings. The procedure also assumes some practical and legitimate method of breaking down the unity of a given decoding (the attempt to make coherent sense of the text) into formally discrete aspects which could be differentially attributed to appropriate social variables. One can thus see how one would be led inexorably back towards the system of cross-correlations between demographic variables and broadly categorised

236

(that is coded) responses, typical of the opinion survey – albeit now more sophisticated in its sociological theory of the audience and its ideological analysis of decodings.

We would argue that the above is very close to what the empirical study of *The 'Nationwide' Audience* does. We would see nothing intrinsically 'wrong' in such an approach if this is indeed what you want to do. The point is, however, that the whole of Morley's theoretical account suggests that his objective (an 'ethnography of reading') is something quite different from or more elaborate than what he characterises as 'a fairly banal exercise' of recording 'correlations and patterns'.[8]

In fact, it is not a case of 'either/or' but the tension between the two approaches that structures Morley's study, a tension that is apparent in the status and treatments of the audience groups. On the one hand, the groups are reported to be real social entities with an existence independent of the research and they frequently exhibit an unstructured, empirical mix of gender, age and ethnic backgrounds. On the other hand, every group is formally categorised as possessing, and analysed as a function of, a homogeneous class identity; while Morley points out that he 'attempted to construct a sample of groups' which would be likely to represent a range of different decodings.[9]

From this point of view, the real complexity of each group and of its collective decodings of the text is not the object to be investigated but a problem to be circumvented. Differences in the sociological characteristics and decodings of individuals within any group are presented as inessential factors to be stripped away in order to reveal their common identity '... subsidiary differences ... which are to be acknowledged but which, I would argue, do not erase the patterns of consistency and similarity of perspective within groups.'[10] The groups thus effectively operate not as real social entities but as convenient and non-essential methodological devices. Each group is little more than an economical way of formally aggregating individuals with a common class position in order to reveal the common decoding corresponding to this beneath the empirical diversity of individual responses. We can see absolutely no reason (other than economical use of resources) why this could not have been done by interviewing the individuals separately.

The same pattern is visible in the way the problem of context is handled. One would expect any materialist study of how different social groups decode television to pay particular attention to contextual questions – from the groups' immediate conditions of existence to the broad socio-historical and cultural context. Morley's groups, however, are treated less as real participants in a specific socio-historical environment than as research respondents who merely formally *represent* the 'genuine' social groups existing elsewhere and at a more fundamental level (that is, class).

It is, in fact, neither possible nor necessary for the survey model of research to tackle contextual questions adequately. Respondents inhabit a largely technical, abstract space in which context only intervenes as a possible external source of interference with the unconstrained response of the individual. Firstly, then, Morley radically reduces context from being the material conditions of existence of social agents to what he calls 'situational variables' in the immediate context of the decoding and interviews.[11] Secondly, these situational variables are only significant if they do, indeed, vary from one respondent to the next. Now all of Morley's groups exist within the educational system as classes of students following various courses (it is in fact the relative vocational or cultural specificity of the courses which accounts for the class homogeneity of each group). Since all groups are interviewed in 'the same' educational context, however, his is a situational constant which can formally be ignored. It is thus only in anticipating similar audience research being done in a different environment that Morley admits the relevance of contextual factors, and then only in this reduced sense of situational variables.[12]

It is this tension between the real or merely formal status of the decoding groups that Wren-Lewis is gesturing towards in the arguments discussed earlier. Recognising only the survey model of research, however, he sees the group interviews as unnecessary or illegitimate. His call for a return to the concrete individual thus threatens to reduce the (ambiguous) complexity of Morley's research, by removing the ethnographic dimensions in which the groups have, potentially, a quite different status.

In this he appears to have Morley's own support. In his retrospective evaluation of the *Nationwide* study, Morley recognises the problem of the groups' empirical or formal status in terms of '. . . the relation of empirically observable groups to the concept of class: within the study the groups are referred to in such a way as to grant them implicitly a representative status – they are taken to stand for segments of society – in this case, classes.' His conclusion, however, is only that his study failed to meet the norms of scientificity required by the representative survey: 'Minimally, given the small size of the sample studied, there is a problem about generalising the conclusions of the study in any way which took for granted the representative nature of the groups – the groups could only be taken to have a potentially illustrative function.'[13]

What we would argue, nonetheless, is that in terms of the research project and objectives developed in the opening theoretical sections of *The 'Nationwide' Audience*, it is not the decision to work with groups that is at fault, nor the rigour of the representative sample which is at stake. It is rather that the use of the survey model itself compromises the ethnographic, qualitative and contextually specific aspects of the research by radically abstracting from the real material complexity of the groups.

This entanglement of the two approaches is thus also visible in an apparent contradictoriness in the relation between the hypotheses, research and conclusions of the study. As we have seen, the groups effectively represent class positions. What the study is thus able to compare is the relationship between the different class positions and the different decodings of the groups. This is in line with the first hypothesis of the research that 'decodings might be expected to vary with 1) *Basic socio-demographic variables*'.[14] The conclusion drawn from the research, however, is that 'social position in no way directly correlates with decodings'.[15]

This failure to find a correlation does not of course disqualify the research in any sense – negative results are every bit as important as positive ones. Indeed, this absence of any correlation *confirms* one of the research's premises which appears at least superficially at odds with its own hypothesis – the critique of what Morley calls 'sociologism' which 'attempts to immediately convert social categories (e.g. class) into meanings (e.g. ideological positions) without attention to the specific factors governing this conversion'.

The trouble here is that by stripping down the real social complexity of the groups in favour of studying correlations with a formal variable, the research is not effectively empowered to demonstrate *more* than this absence of a correlation. The influence of a largely inappropriate research model thus deflects the research from its *primary* objectives detailed in the statement of hypotheses cited above:

> decodings might be expected to vary with ... 2) *Involvement in various forms of cultural frameworks and identifications*, either at the level of formal structures and institutions ... or at an informal level in terms of involvement in different sub-cultures ... Given a rejection of forms of mechanistic determination, it is at this second level that the main concerns are focussed.[16]

What Morley actually seeks to demonstrate is that there *is* in fact a material relationship (and not just a formal correlation) between social position and decodings but not a direct, mechanical one. To do this he has to demonstrate that (a) decodings have some regular connection with institutional or subcultural factors (the proper level of ethnographic study); (b) these factors themselves have some material connection with broader social categories like class; (c) decodings do not directly correspond to these categories. In the relative absence of genuine ethnographic analysis, he can only demonstrate (c) effectively, which on its own is simply negative, inconclusive and unable to support the arguments of (a) and (b). This is emphatically not to suggest that the ethnographic dimension of (a) is absent from the research, but that it is not systematically realised by a methodology whose concerns lie elsewhere.

We would accept as wholly reasonable Morley's contention that some of the decodings reflected the influence of working-class populism, trade union and Labour Party politics, black youth cultures, etc. We do not feel, however, that the existence of such relationships, yet alone their nature and significance, is systematically and satisfactorily elicited, demonstrated, interpreted or explained in the book. Nor is this likely to emerge from a methodology which treats the decoding groups as decontextualised aggregates of individuals, or whose focus is on representative correlations between one-dimensional class positions and monolithic formal responses to a text (either 'dominant', 'negotiated' or 'oppositional', following Parkin's scheme).

What Do Groups Do?

Rather than emphasising what groups represent, then, it might be better to concentrate on what groups do. This stress on the active, productive role of audiences would seem at least more in tune with the distinctive character of the encoding/decoding model from which Morley's research begins. A possible objection, however, is that by focusing on the group in its own right one thereby loses the connections with the broad social structural factors like class which have, even if indirectly, a determining influence on the way people decode television.

We would argue, on the contrary, that this is only a problem if one insists on thinking of class as a 'thing' exercising some external effect on another 'thing' – the group. The complex structure of social relations within an audience group constitutes the immediate context and material conditions in which decoding takes place and, as uses and gratifications approaches could tell us, one ignores them at one's peril. Despite what uses and gratification approaches might tell us, however, in studying these social relations one is studying nothing other than a moment in the material existence of class. Class is internal to the group; and while its existence extends far beyond, it is always embedded in the material practices of other institutional, subcultural and organisational forms apart from which it could have no empirical existence. It is perhaps true that we more naturally think of class on the 'macro' level – as the structural tendencies objectively installed in the social system as a whole; or as the broad strategic co-ordinations of 'local' social relations, which have the capacity to reproduce or significantly transform this social structure. But it is just as valid and important to understand class at the level of that vast and shifting network of social relations within and between the nodal points of formal and informal institutions, organisations and subcultural groups.

What we would suggest is that Morley's reliance on a representative survey model makes it difficult for him in practice to treat class as anything but a set of *external* determining constraints on the audience – despite his theoretical disavowals of this position. Correlations and relationships whereby a group stands for a class are formal external relations between separate phenomena. Insofar as these are taken to signify the existence of real relations, then, these too appear as an external connection between class 'over there' and the activity of human beings 'over here'. Social determination thus appears as a process acting *on* human beings rather than a human process. It would perhaps be better to think of social determination as Marx's 'Iron Law of Tendency' – a process which is real and operative but only ever so through the actions of human beings which may realise, inflect or completely negate its effects.

The problem emerges visibly in Morley's attempts to define the determining chain 'class – subcultural discourse – decoding' in a non-deterministic way. Clearly, to simply insert the mediating term of subcultural discourse between class and decoding no more changes the mechanistic nature of their relationship than does extending a line of touching snooker balls. Morley therefore argues that what class determines is not which discourse the decoder mobilises in engaging with a television text, but 'the structure of access to difference discourses'.[17] Now this is close to the idea of tendency only realised through conscious human action. But coming at it from the direction of the one way determining effect of class *on* human action and in terms of the opposition of determination *to* human action, there is no way Morley can think through the nature and unity of the human, social processes at work. What he is left with is an area of untheorised, statistical indeterminancy, representing a *formal* balance between equally unreal extremes of arbitrary human action or determinism: 'That such relations are only probabilistic is clear (i.e. it is simply more *likely* that a person in social position X will have access to a particular form of cultural competence than a person in position Y).'[18]

The question 'what does the group do?' avoids some of these difficulties by stressing both the investigation of empirical relations rather than formal ones and the constitutive role of the audience group's activity. To use Morley's favoured metaphor, it involves the study of cultural 'performance' rather than cultural 'competence', what people *actually* do rather than what they *can* do. Decoding is thus not thought of in terms of the formal possibilities and limits of interpretation inherent in the discourses available to a particular group. Turned on its head, the question is the more empirical 'what does this audience group do with the text it chooses to watch and what discourses does it mobilise in the process?' This is also to say that the ideological discourses and social relations which define the group and constitute the immediate conditions in which decoding occurs are not taken as simply given or fixed determinants of the decoding but as conditions which are capable of being changed in and by the decoding process.

This is perhaps the key lesson brought home to us by our own work with audience groups, and the target of an ethnographic approach. The process of negotiating a collective decoding of a text in an established audience group involves it in a considerable work of clarification and renegotiation of its own 'internal' discourses. Ultimately, these changes are also materialised in changes in its own social hierarchies, practices, etc. Such changes, moreover, cannot be understood as simply internal to the group, in that they are necessarily transmitted, via both individuals and group practices, to other, 'neighbouring' institutions and informal organisations. In this sense, then, the process of decoding can also be understood as a contributory moment in the reproduction/transformation of that pattern of ideological, cultural and organisational practices which defines a class at any given point.

One must, of course, be careful not to overestimate the actual extent of these processes and their effects. Firstly, they are heavily concentrated in the group at a discursive level and are likely to achieve more material effects only when repeatedly confirmed – either extensively over a period of time, or intensively by the same changes taking place simultaneously across a range of groups. Secondly, while this corrects a certain one-sided stress on determining conditions in thinking about decoding, even the incremental effects of these 'counter-determinations' are, except in exceptional circumstances, likely to be marginal inflections and resistances to the sheer dead-weight of the existing, broader forms of social organisation. This having been said, these processes are real and observable, at the discursive level and beyond, even in the study of 'one-off' decodings. They therefore require taking into account if only in order to understand the process of textual decoding and its results.

This requires that decoding is treated, theoretically and in practice, as a genuinely social process. That is to say, the collective decoding arrived at by the group is necessarily different from that which any individual member could have produced. This is because the group decoding is constructed through *a series of interactions* between a progressively modified reading of the text and the progressively renegotiated identification and definition of the appropriate discourses through which it is to be read. It is here that the ultimate justification for using group interviews lies, and, indeed, something approaching this does form part of Morley's declared motive for using groups:

> The aim was to discover how interpretations were collectively constructed through talk and the interchange between respondents in the group situation – rather than to treat individuals as the autonomous repositories of a fixed set of individual 'opinions' isolated from their social context.[19]

Nonetheless, this aim is not, we feel, fully followed through in the way he presents the empirical record of the group decoding process, where it is again compromised by the pressures of a methodological formalism. What tends to be demonstrated is very much 'decoding' in the sense of a finished product and the component parts of that product, rather than 'decoding' as a process and the substantive ideological shifts which mark the stages of that process. In practical terms, this would require a much fuller presentation of the decoding process for each group (but then the need for a representative range of groups to be discussed would not be pressing); and the identification of each individual's contributions to discussions in order to be able to follow the negotiated shifts of ideological position/reading. In theoretical terms, it requires the replacement of formalist conceptualisations of decoding as firstly, a simple reader/text relations, and secondly, a formal relation to 'fit' or 'mis-match' between decoding and preferred reading.

How should one define the decoding process? One response is to follow the naturalistic argument of Wren-Lewis in restricting the moment of decoding to the actual viewing and accompanying reading of the television text – necessarily conducted by the individual even if in the company of others. Any encounter with other interpretations, commentaries, discourses either during or subsequent to this process is thus regarded as a related but different process – the general ideological discussion, elaboration, modification of an already completed decoding. Decoding is thus established as a tightly enclosed circle of text and reader in which the preferred reading of the text and the reader's cultural competence converge to define 'the' final meaning of the text. Such a view is already implicit in the artificial terminological symmetry of the 'encoding/decoding' description. Despite the fact that the interest in and value of this model is that it restores some active, constitutive role to the audience, this seems a singularly 'inactive' or unproductive form of activity since it only leads back to texts. What do audiences do? Watch television.

Alternatively, since the decoding made by any individual is already a socially formed reading, one could reject this naturalism and regard decoding as a group activity. Decoding now appears as a more extensive, indefinite process – even perhaps renewable in different group contexts. If it could be said to have a definite endpoint (rather than simply stopping), this would be the arrival at a group consensus – or at least an agreement to differ – on the nature of the decoding. Decoding now also involves the collective discussion of more general ideological positions, while remaining identifiably a process of decoding through its repeated return to the text as the common reference point, agenda and target of that discussion.

This extended definition of decoding involves the more productive activity of a collective 'appropriation of meaning' – more productive since it leads outwards to the uses made of televisual representations and hence to their immediate practical effects. In this sense, decoding is not a simple text-reader relation but two simultaneous and interacting series of negotiations – progressively defining and reconstructing the meaning of the text through collective negotiation of individual 'part readings'; while progressively redefining the nature and unity of the group through the collective negotiation of a textual decoding. To draw illustrations from our own research, we could cite the group of college catering students where the decoding of a current affairs text was only constructed through the establishment of a hierarchy of importance between their collective self-identification as young workers or as students – with consequent implications, also discussed, for the kinds of organisational and practical involvements they entered into with other groups of students or workers. Or again, the case of a radically heterogeneous group of 'the unemployed' where the struggle to arrive at a consensus decoding of the text was simultaneously a process of recasting the conventional ideological and social relations inherited by the group and exploring the often unexpected alliances, hierarchies and relationships which might define and structure a subculture of the unemployed.

In all this, there is of course a danger of falling in with what seems to be a current, populist revival of uses and gratifications approaches. Here the role of the television text is wholly neglected in favour of generally unstructured accounts of 'television talk', of what people do with television. Firstly, then, it is important to reaffirm that the material with which the groups engage through the kinds of negotiations illustrated above is not simply 'television' or 'programmes' but a specific text. That is to say, it is with a structured system of representations which only retain their relative ideological coherence if read in one of a limited number of ways (the 'preferred reading'). To be sure, in decoding a specific text it is possible and common for audiences to fragment and restructure this ideological coherence to a greater or lesser extent. But this is always at the cost of problematising the text itself, of opening up contradictions and incoherencies between textual elements where before there was only seamless continuity, of creating lacunae where representations not present in the text are required to sustain the ideological coherence of the decoding, of rendering elements of the text redundant, etc.

It is generally the effort to account for these contradictions, gaps, redundancies – *perceived as features of the text* – which powers the decoding group's discussion and drives it further afield. In so doing, it can begin to reveal (to the group as well as the researcher) the real source of these 'textual' discrepancies in discrepancies between the dominant codes of the text and those effectively but often unconsciously operating in the group. As a result, the group begins to articulate, define, refine and consciously reappropriate its 'own' ideological and cultural discourses, with the further potential effects noted earlier.

The general point here is that the audience's (always incomplete, unfinished) reappropriation and redefinition of its own discourse and identity is *only* achieved through the struggle to appropriate and redefine a structured text. This process can therefore hardly be adequately analysed without adequate understanding of and reference to the available textual materials through and on which it works. Beyond this, we cannot do better than, like Morley, quote Steve Neale's view of what is required if what people do with television is to become properly intelligible: 'What has to be identified is the *use* to which a particular text is put ... within a particular conjuncture, in particular institutional spaces, and in relation to particular audiences.'[20]

Nonetheless, although Morley upholds in principle this stress on the particular and specific, in practice it is undermined by his insistence on using Parkin's tripartite model of 'dominant', 'negotiated' and 'oppositional' readings. This is despite his prior recognition of its inadequacy as 'a logical rather than a sociological statement of the problem ... providing us with the notion that a given section of the audience either shares, partly shares, or does not share the dominant code in which the message has been transmitted'.[21] The continued use of the schema is consistent with the way the actual research of *The 'Nationwide' Audience* is designed to investigate the existence of formal correlations between preferred meaning, decoding and class position. The formalism of the schema, however, is not adequate to the more ethnographic and extended conception of decoding, where it is the substantive ideological content of the decoding which is of interest.

A marginally better recasting of the schema would be to view *all* decoding as negotiation with a preferred reading, though the effects of that negotiation might be to broadly reproduce, inflect or negate it. By putting all decodings on the same level rather than allocating them to mutually exclusive pigeon-holes, this at least has the following advantages. It requires, firstly, that one defines the substantive ideological relationship between text and decoding – *how* the text is appropriated rather than whether there is a 'fit' or not. Secondly, it enables one to handle the vast empirical diversity of possible relations between decodings and preferred readings without an endless taxonomic proliferation of subcategories to formally classify them, which is the solution recommended by Morley.[22] Thirdly, and more particularly, it allows one to think through the potential for ideological resistance contained in any and all negotiations with dominant codes.

For Morley, this resistance tends to be identified with Parkin's 'oppositional' category and thought of largely as an absolute refusal of the text's preferred reading – either by the substitution of a relatively coherent, 'radical' alternative or by simple non-engagement with the text, as with his group of black students. We would suggest that this resistance to dominant codes is more commonly carried and expressed within those codes and discourses themselves. It exists less in the form of cogently articulated, systematic discourses brought to bear on the text from outside; but rather in the practical effects of a largely implicit level of discourse concealed/expressed within inflections of the language, representations and coherence of the dominant codes themselves and only articulated in a fragmentary way. What we have referred to as the group's conscious reappropriation of its own discourse is essentially the articulation of these implicit resistances in the discursive forms necessary to their conscious and practical use.

For example, in the decoding of current affairs television it is in our experience common for the effective if inconsistently sustained presence of an oppositional discourse to be signalled and expressed through the vocabulary of 'bias'. Brought to light and elaborated, this discourse evokes quite radical challenges to the professional authority, practices and ideologies of broadcasters, to dominant representations of 'normal' and 'legitimate' politics, and so on. However, since 'bias' is defined within the dominant codes of current affairs television in terms of impartiality, balance, even-handedness, etc., and broadcasting is generally irreproachable at this level, it is easy for researcher and decoding group – *but particularly for the individual decoder* – to miss the presence of this discourse or recognise only an 'inaccurate' decoding of the text.

We would see a major objective of ethnographic audience research as the eliciting and identification of this level of discourse and resistance. For it is here that one finds revealed the *differentia specifica* which inflects decodings from the preferred reading of the text. Morley's discussion of the group's 'structure of access to discourses', is, in this respect, potentially misleading. It may be taken to imply that the discourses a group brings to bear on the text are simply ready-made or explicit; or that these discourses are already the group's own, that their employment by the group does not also involve it (to a greater or lesser extent) in ideological struggle with these discourses as well as with those of the text. A text, for example, whose preferred reading involves the condemnation of strikes may be read by a group through 'the' discourse of trade unionism. But one cannot anticipate the decoding that results, whether it will broadly reproduce or challenge the preferred reading, insofar as trade union discourse is itself the contradictory articulation of a dominant 'responsible' trade unionism with a subordinate and less formalised 'wild' trade unionism. The processes of decoding we have been discussing here would involve the group in 'placing' itself ideologically and hence practically in the discourses and practices of trade unionism.

Finally, this eliciting of a partially submerged and fragmentary level of discourse has a number of practical methodological implications which are worth mentioning. Firstly, it turns to advantage the 'problem' that decoding groups in a research context tend to produce more critical readings of texts than might otherwise be the case. While we would emphasise the need to investigate real social groups with an existence independent of the research, this is not however to condone the naturalistic illusion that the research enterprise does not intervene in that reality. What one is engaged in in this case is merely intensifying and focusing on existing processes.

Secondly, *any* academic research involving interviews with respondents will tend to 'interpellate' them as responsible, serious, official subjects and elicit responses framed within the respondent's perception of official discourse. This is, however, clearly a particular problem here. Working with independently existing groups with a sense of collective identity and solidarity mitigates the difficulty, as does use of the informal, non-directive interview, outlined by Morley, where the group sets the agenda, frames its discussion in its own language, etc., and where the interviewer's role is largely restricted to seeking elaborations and clarifications of meaning. It would also seem important to interview the group on its own ground wherever possible. Significantly for *The 'Nationwide' Audience*, however, our own experience suggests it may be almost impossible to escape the double confirmation of an academic role involved in the use of an educational context.

Conclusion – a 'Uses and Effects' Approach?
We would stress again the need to study the decoding of television in terms of real groups operating in specific material conditions, rather

than in terms of formally representative entities. We therefore readily acknowledge that the value of ethnographic audience study and its extended conception of decoding is radically reduced if it is not the case that people commonly use television in the ways we have discussed. Now it would be hard to deny that for much, if not most, of the receiving of television, decoding is a process that ends with the individual's reading and interpretation of the text.

Equally, however, there are important areas where the decoding of texts is accompanied by a commentary and discussion through which viewers negotiate their ideological and social relations with one another. The most widespread and obvious example would be the family; although, in both its socio-cultural characteristics and its relation to television, it is something of a special case compared with other institutional and sub-cultural forms. A second area where decoding already exists as an extended process could be categorised in terms of particular textual genres. Again, the most familiar example would be soap operas and melodramas. Their quasi-real status as something like an experimental social simulation fosters and almost requires the continually renewed intervention between episodes of a public debate of the text and its moral messages – a process confirmed and complicated by contributions to that debate by the media themselves.

A third set of examples might be found where the collective ethos or practices of a sub-culture or institution are commonly carried over into the group-viewing and discussion of relevant programmes: pop music and youth audiences, television coverage of football, etc. Finally (though the list is not meant to be exhaustive) there are those cases, most easily associated with current affairs television, where specific issues or events are felt by particular audiences to be sufficiently important to their interests to call 'naturally' for collective viewing and evaluation of their representations on television – often over a relatively extensive period: key elections, the more intense social, industrial and political struggles, moments of national crisis, etc.

The choice between a restricted or extended conception of decoding the individual or group respondent, then, is not a simple matter of definition but part of the whole way one sets up the general parameters and objectives of audience study. It might be argued that the examples of decoding given above are all special cases and that the findings of any study which concentrates on them would thus not be representative of the more restricted and mundane character of decoding in general. In our view, however, while the results may not be *representative*, they are likely to be more *typical* of the general nature of decoding, as well as (and because) more illuminating. What are revealed in the extended, group processes of decoding are the same (typical) laws and processes which govern the restricted decoding of the individual viewer, but more fully realised and thus more visible and intelligible as social processes.

This is also to say that it does not strike us as *all* that valuable to know how and why Mr Brown and Ms Smith decoded the same text differently, unless we also know *what difference it makes*. The effects of that difference will be present in the individual's marginal ideological restructuring which will also have the latent potential to affect the reproduction of the formal and informal organisations in which she/he participates. But it hardly needs saying that it is almost impossible to discern meaningful effects at this level of particularity. We have tried to suggest one way in which it *is* possible to begin to discern these effects in cases like those listed above – where, of course, 'effects' are to be understood as the ideological and social changes actively brought about by audiences, in the process of using and adapting television representations to rethink and reorganise their own conditions of existence.

References

1. C. Brunsdon, D. Morley, *Everyday Television: 'Nationwide'* (London: BFI, 1978); David Morley, *The 'Nationwide' Audience* (London: BFI, 1980).
2. Morley, *The 'Nationwide' Audience*, p. 33.
3. J. Wren-Lewis, 'The Encoding/Decoding Model: Criticisms and Redevelopments for Research on Decoding', in *Media, Culture and Society*, vol. 5, 1983.
4. Ibid.
5. D. Morley, 'The "Nationwide" Audience: A Critical Postscript', *Screen Education*, no. 39, 1981.
6. Wren-Lewis, 'The Encoding/Decoding Model', p. 195.
7. Ibid., p. 194.
8. Morley, *The 'Nationwide' Audience*, p. 17.
9. Ibid., p. 26.
10. Ibid., p. 138.
11. Ibid., p. 27.
12. Ibid.
13. Morley, 'The "Nationwide" Audience: A Critical Postscript', p. 9.
14. Morley, *The 'Nationwide' Audience*, p. 26.
15. Ibid., p. 137.
16. Ibid., p. 19.
17. Ibid., p. 134.
18. Morley, 'The "Nationwide" Audience: A Critical Postscript', p. 9.
19. Morley, *The 'Nationwide' Audience*, p. 33.
20. S. Neale, 'Propaganda', in *Screen*, vol. 18 no. 3, Autumn 1977; Morley, *'Nationwide' Audience*, p. 18.
21. Morley, *'Nationwide' Audience*, p. 20.
22. Morley, *'Nationwide' Audience*, pp. 20–21.

HANS VERSTRAETEN

Commercial Television and Public Broadcasting in Belgium: The Tension Between The Economic and the Political Dimension

The Politico-Economic Reference Frame

The current conflict between commercial and public broadcasting in Belgium is clearly no isolated phenomenon, but part of the international trend towards an ever-increasing commercialisation of the media. In order to fully understand the dynamics of recent media developments, it is necessary to situate these within a general social evolution. The following three premises seem to me crucial for an adequate analysis of current international media questions.

The Increasing Expansion of Exchange Value

Social evolution since the origins of capitalism has been characterised by an unremitting expansion of the domain of exchange value. The economic principle of profit is widening more and more and slowly absorbing the whole sphere of people's lives. Social sectors which could previously evolve in relative autonomy from the profit principle (for example, physical and psychic health, leisure, education), are becoming ever more subjected to exchange value. This has also occurred in the field of the media. The components of the 'ideological apparatus', among them the media, have gained an economic dynamic of their own under late capitalism. This was understood a long time ago by the Frankfurt School in its discussion of the 'industrialisation of culture'. More important, however, than merely establishing this 'expansion of exchange value', is its politico-economic analysis and evaluation. For the industrialisation of culture is not an unproblematic, monolithic process, as the Frankfurt School regularly tended to regard it, but instead leads to a number of areas of tension between the politico-ideological and the economic-social domains. For instance, the commercialisation of the daily press (parallel to the rise of the mass circulation press at the end of the 19th century) led to the demise of the party-political press. For developments in the field of the economic infrastructure do not always automatically coincide with prevailing ideological interests, but often occur through a process of transformation marked by conflict. It is my purpose to describe in this essay how this field of tension between the political and economic dimension has taken shape in the specifically Belgian context, with reference to the dismantling of the public broadcasting monopoly and the implementation of a commercial television station. My analysis on the whole concentrates on the situation in Flanders.

The New Media and Their Economic Functions

The problem of commercial television cannot be separated from other developments at the level of the 'new media'. For analytic purposes it is necessary to differentiate the new media into two groups. On the one hand there are those 'new' media forms, which in essence only offer a new distribution channel to a traditional media content (cable television, pay-television, satellite television, video, 3-dimensional TV). On the other there are those new media technologies which carry a new media content: teletext, viewdata, data networks, teleconferences, microcomputers. This second group owes its development mainly to an integration of informatics and telecommunications.

However, these two kinds of new media do not only differ in content. The main distinction is situated on the level of the difference in economic functions. The first group of new media, which for ease I will refer to as 'first generation' new media, has a dual exchange value function, as is also the case with the traditional media. There is in the first place the function of accumulation with respect to media capital itself, in the second the function of circulation with regard to the total social capital (especially through advertising).[1] These first generation new media therefore find their origins in the necessity for safeguarding the accumulation of the media industry's capital. In contrast, the 'second generation' new media find their origins in the objective need, not so much of media capital itself, but of the total social capital. Informatics and telematics are in the main an answer to the demands made by the shift in production relations and find their social origins in, among others, attempts to rationalise the tertiary sector (the applications were later extended to the production sector itself). Hence a number of important conclusions are to be made with regard to the diversification of the new media of the second generation versus those of the first generation.

Naturally these second generation new media are also subject to a double exchange value in their functions of accumulation and circulation, but the hierarchy between the two economic functions here is quite the opposite. In contrast to the new media of the first generation, the function of accumulation is in this case subordinated to the function of circulation with respect to the total social capital.[2] The

252

first generation new media, it is true, do generate an acceleration of capital circulation, but do not as such affect the production process in itself: the proportion between labour and capital in the production process remains constant. The situation is totally different with the second generation new media. For these directly affect the production process itself. Not only are labour relations fundamentally altered by this, but these new communication technologies are moreover employed to implement a rationalisation of the constant capital.[3] The new media therefore have an economic function which not only has a clearer profile, but is also different qualitatively from that of the first generation new media.

Even though the use of these second generation new media is increasingly widespread, there is still talk, in the field of application within the production system, of a low rate of growth of productivity in information handling, and lack of innovation. The finest technology is already in existence, but for the time being is being used on a rather limited scale. The president of Communications Studies and Planning Inc., an international research and consultancy organisation in telecommunications and information technology, is very clear on this matter: 'Technologies that can yield very large savings in cost and staff time – teleconferencing, electronic mail, and advanced methods of document storage and retrieval, for example – have been available for at least ten years but have so far been implemented only on a very small, though increasing, scale'.[4] The so-called information society, based on these second generation new media, has for the time being been only partly achieved.

The fact that these new media are, in the main, designed for the purpose of rationalising the production process, nevertheless has a drawback for the media industry itself. These technologies, and related services, are principally attuned to the needs of producers and much less to those of consumers. Indeed, this new information technology is having difficulties in penetrating the consumer market (note, for example, the difficulties in elevating the viewdata services above purely professional use). The accumulation function of media capital itself is thereby slowed down. It is therefore of prime importance that these second generation media, if they wish to further expand the accumulation of their respective capitals, succeed in finding a larger outlet, not only in the producers' market, but also and especially in the consumers' market.

Integration of the Different New Media
The above distinction between the new media of the first and the second generation was made on purely analytic grounds. In social reality these two kinds of media forms can hardly be separated from one another. On the contrary, the main driving force behind new

253

media developments is vested precisely in the technological possibility of interconnecting these two different media forms (via cable, satellite and ISDN networks). The information society of the future will only be possible when use is made of a strategy combining these two media in an appropriate manner. The first generation new media (cable, satellite and pay-television, video) aim mainly at bringing entertainment programmes. The hope is to install, for this quantitatively enlarged entertainment offer, the transmission infrastructure (see the plans for cable and satellite), which can then be used for the new media of the second generation, that chiefly carry information. A recent study rightly states: 'Although the most apparent and immediate drive to cable television systems has been the perception of a market for new entertainment services, the growing corporate interest in cable is clearly related to the wider telecommunications issues'.[5] The current projects extending the various cable television services, satellite television and pay-TV are merely a preparatory phase, which, however, is necessary to persuade government to lay the necessary transmission infrastructure. The first generation new media hereby serve as a pioneer to make it possible to bring the second generation new media into the living room. At the same time this explains why, with the help of banks and other financial concerns, international media conglomerates, active in both fields, have emerged in the past few years. It seems to me most important to keep in mind this 'combination strategy' when studying the problem of commercial television. An analysis of an 'enlarged' television supply within the specific context of a given country must not be isolated from the underlying structural dynamics which to a large extent condition it.

Commercial Television in Flanders: The Tension Between Politics and Economics

Contextual Factors

In the past few months the discussion seems to have taken off in Belgium regarding the dismantling of the monopoly public broadcasting position with the creation of a commercial station. [Since this article was written in 1986, commercial broadcasting has been introduced. – Eds.] Publications, political positioning and conferences have succeeded each other at an ever-increasing pace. There are four central contexts.

First, one should note the rapid and important developments, not merely in the technological field, but especially in the domain of financial interests and the commercial strategy of both the national and multinational media groups. The plea for the privatisation of

254

Belgian state companies emanating from financial circles, and the creation of Media International by Albert Frère and Rupert Murdoch, are prime examples of this.

A second factor is the lack of a coherent government media policy. As far back as December 1981 the government made known its wish to abolish the public broadcasting service's monopoly position. Nearly five years later opinion within the majority parties is still divided on the best means of realising this in practice. The political infighting between Christian-Democrats and Liberals during the last coalition talks illustrates this.

Third, there is the inability of the political opposition to present an alternative media blueprint. The organised left in Belgium has always paid little attention to media problems.[6] Left-wing points of view regarding the media usually show defensive and pragmatic traits. Their defence of public broadcasting is as yet far too steeped in a sacrilised view of the public broadcasting service. Due to the lack of long-term vision in relation to media policy, the political left is often driven to implement rather inconsequential turnabouts in its standpoints regarding the media (for example, regarding the introduction or not of advertising on the public broadcasting service).

A fourth element is the preponderance of ideological and commercial interests on the part of the players concerned. This leads to a media debate insufficiently supported by a scientifically grounded analysis of facts, for instance in the assaults upon the news service's lack of objectivity. From commercial circles an appeal is made to the presumed wish of the public to carry through the commercialisation of the media scene.

Within this field of tension, created by the four above-mentioned factors, the pressure to start a commercial television service has become even greater. However, commercial television was cherished as a wishful dream in Belgium during the sixties. This wish was originally uttered in party-political surroundings, from where strong criticism was levelled at the so-called lack of objectivity of the news service. This criticism was further developed in the daily press, which in Belgium is closely allied to the party-political groupings. It can be shown through content analysis that criticism of the news service's objectivity in fact acted as the political inducement and motive for juridicially breaking the monopoly of the public broadcaster.[7] But the principle of dismantling the broadcasting monopoly was not only supported in party-political surroundings, but by commercial private industry as well. The Belgian advertising industry has, for some twenty years, been clamouring for permission to broadcast their adverts on Belgian television. As long as the political battle remained to be fought to make government accept the principle of breaking the broadcasting monopoly, the interests of the party-political and private-economic circles converged nicely. As soon as the government, in its policy declaration of December 1981, accepted the principle of dismantling the broadcasting monopoly, its practical application had to be worked out. It was precisely in this area that the party-politicians' and private industry's interests were much harder to harmonise. From a party-political angle a media model was envisaged whereby the broadcasting monopoly would be broken by a commercial broadcasting station, equipped with a substantial news service (as an alternative to the public news service), financed through advertising and with an important participation by press groupings. Nevertheless, it was soon clear that this model was not entirely appropriate for private industry; before throwing in their lot with the adventure, the press groups commissioned a feasibility study which made it obvious that at best there was room in Flanders for only one commercial television station, thus creating a number of problems for private industry.[8]

The Discord Within the Flemish Press
The negotiations between the Flemish press groups to present a common project for commercial television proceeded with great difficulty. This led to a schism within the Flemish press in June 1984. Almost at the same time two press groups were formed, which both wanted to become involved with commercial television: OTV (Onafhankelijke Televisie Vlaanderen – Independent Television Flanders) consisting of the paper groups VUM, Concentra and the Financieel Economische Tijd and two financial groups (Electrafina, Gevaert Photo Producten); and VMM (Vlaamse Media Maatschappij – Flemish Media Company) grouping together all other Flemish newspapers and weeklies. These two groups are marked by important differences, as illustrated in Table I.

TABLE I
Differences Between OTV and VMM

Differences	OTV	VMM
1. political	non-pluralistic (only Christian-Democratic press)	pluralistic (press groups with various political affinities)
2. financial	input from financial groups	no input from financial groups
3. economic structure	allied with international media groups	Flemish press groups only
4. commercial strategy	offensive (geared to second generation media)	defensive (commercial television only)

256

It is clear from Table I that in merely economic terms, the OTV group offers the most interesting perspectives. First, it manages a broader capital base, thanks to the input of the financial groups. Electrafina and Gevaert participate directly in OTV, while the group may indirectly also expect financial support from the three largest Belgian holdings (Groep Brussel Lambert, Cobepa and Generale Maatschappij). Second, OTV is part of a financial-economic network that has steadily diversified its media activities in the past few years (telecommunications, data transmission, viewdata). So commercial television is only part of a much larger plan for media exploitation, clearly geared, additionally, to the media of the second generation. Third, OTV has several connections with international media groups (RTL, News International), and therefore has an important lead in the area of know-how and software.

Notwithstanding its commercially favourable business structure, the OTV project is difficult to realise within the Flemish political context. The breaking up of the broadcasting monopoly must be elaborated by government and approved by parliament and the chance of finding a political majority for OTV is very slim, for two reasons. First, the OTV group is not constituted pluralistically; its principal elements are the Christian-Democratic press groups, and the Liberal government coalition partner is not represented. Second, the OTV group from the start had indirect financial ties with the Luxembourg commercial broadcaster RTL, via the Electrafina holding company.[9] Acceptance of the OTV project therefore entails the inclusion of a foreign media group, making the project politically unviable.

This illustrates the area of tension between the political and economic dimension, created by the ongoing commercialisation of the media. The OTV project, the most interesting on economic grounds, is nevertheless difficult to achieve at this moment for political reasons. The VMM project on the other hand, commercially and economically offering less interesting perspectives, has, for the same political reasons, a better chance at present.

The Economic Risk of a Commercial Station
It is most unlikely that a commercial television station as envisaged by the party-politicians can be operated on a commercially sound basis. That commercial television abroad (Italy, Great Britain) turns out to be a financially successful formula, does not mean that this would also be the case in Flanders, where the media context is fundamentally different in a number of important ways.

First, the potential market for a Flemish commercial broadcaster is particularly small. The viewing audience totals 5 million people at the most, of which the commercial station will naturally only be able

257

to capture a segment. The expected advertising revenue for a commercial station is estimated at BF 2.5 billion (about £38 million) annually at the highest (and this only after a period of 4–5 years). By comparison, the working budget of the Flemish public broadcasting service for 1985 totalled BF 5.5 billion (£80 million).

Second, as is commonly known, Belgium is a nation with an exceptionally high cable penetration; at present 85% of television owners are connected to a cable network. This makes Belgium outstandingly attractive for international satellite television. Such commercial satellite television projects are now in hand, run by an international media group, and threaten in future to erode the already meagre advertising revenue of a Flemish commercial station even further.

Third, a considerable proportion of the advertising revenue of the commercial station will have to be handed over to the press, in compensation for the loss of press advertising. This point of view has for years been propounded by the newspaper publishers,[10] and accepted by almost all political parties, even though recent analysis has proved that, in the specific Flemish media context, a commercial station entails no real threat to the advertising revenue of the press.[11]

Fourth, the Flemish viewing public is already able to receive fourteen to sixteen different television stations. The possibilities for choice are now already saturated, so that the Flemish audience has no need for an extra television station. Nor are there signs that the Flemish audience would feel a real need for alternative news provision, as desired by the party-politicians: the information programmes of the public broadcasting service gain very high viewing and appreciation figures.

Fifth, the advertising agencies themselves would prefer to see their advertisements broadcast on the public broadcasting service, rather than by a commercial station still to be created. The public channel already serves a large viewing audience and offers the advertising industry the chance of practising audience segmentation. The commercial channel still has to try and find its place in the market and this will in any case lead to a fragmentation of the audience (which severely obstructs selective audience segmentation). Thus a commercial channel, solely directed at a Flemish viewing audience (as envisaged by the politicians), is a very risky undertaking. Nonetheless, for political reasons this model will have to be used in order to create the legal framework for dismantling the broadcasting monopoly. So here we find a concrete field of tension between the economic and political sphere.

New Synergy on the National and International Level
The developments in the matter of commercial television in Flanders must also, however, be connected with a number of financial and

economic regroupings and restructurings that have taken place in Belgium during the past few years. Although it is not possible to examine this in detail in this essay, even a temporary analysis shows how Belgian financial groups are preparing themselves strategically for the information society of tomorrow.

On the national level an important alliance was created in 1982 between the groups Cobepa (with interests in VUM), Brussel Lambert (with interests in RTL) and Frère-Bourgeois, when the former steel tycoon Albert Frère managed to acquire a dominant position in Cobepa as well as the Groep Brussel Lambert, by an injection of new capital.[12] The synergy thus created was especially important with regard to the media because Albert Frère thereby obtained a firm footing in CLT-RTL and a close relationship developed between VUM (which is involved in the OTV project) and RTL.

This synergy, of which the OTV project is part, gained an important international dimension in September 1985, when an alliance came into existence between the Groep Brussel Lambert (A. Frère) and News International (Rupert Murdoch). Together, Frère and Murdoch have created Media International to develop common strategies for the European television market, with particular reference to satellite. Thus, the OTV project is absorbed by a financial-economic network that to a large extent oversteps national boundaries. This alliance between Frère and Murdoch creates an international media network[13] with branches in the USA, Canada, Australia, Great Britain, Switzerland, Belgium and Luxembourg. It will therefore come as no surprise that, a few months after the founding of Media International, the OTV group let it be known that, under the circumstances, it was no longer interested in commercial television for Flanders.[14] It would appear that André Leysen (of VUM), Albert Frère and Rupert Murdoch have other ambitions than a Flemish commercial station, rather limited in scope.

The two synergies outlined show that the strategic decisions concerning the commercialisation of the media are made in the first place by financial interests and industrial groups. Contrary to what is envisaged and wished for by party-politicians, the traditional press groups play a role of minor importance in this. The model concerning commercial television, as the politicians see it (namely, a commercial station of a specific 'Flemish' nature, operated mainly by the Flemish press) threatens, under pressure from this financial-economic restructuring, to be very quickly overtaken by the facts.

The Strategy of the Public Broadcasting Service
It is obvious that the strategies concerning commercial television entail far-reaching implications for the position of public broadcast-

ing, a crisis of the public broadcasting service. However, it is noticeable that in most conceptualisations of this crisis, of its causes and possible relevant remedies, no mention is made of the social context within which it is occurring.

The Crisis of Public Broadcasting: The Causes

It is often asserted in the manifold discussions regarding broadcasting that the public broadcasting service monopoly has lost its *raison d'être* owing to rapid technological developments in the field of the media. The emergence of new and more extensive transmission modes should, following this conception, lie at the root of the crisis of public broadcasting. This results logically from the idea that the public broadcasting service, in its origins, was invested with a monopoly position because of a technological necessity, namely, the availability of broadcast frequencies. Just as there are a good number of indications that seriously call into question a technological-deterministic explanation for the creation of the broadcasting monopoly,[15] so too it seems rather dubious to claim that the crisis of public broadcasting should be brought about by technological innovations. The developments of new communication techniques have undoubtedly played an important role in this, but they do not form the real cause of the crisis of public broadcasting. This crisis dates from much earlier and finds its origins in a number of fundamental problems on the level of the politico-economic structure. I wish briefly to enumerate a few elements, especially since these have almost completely been neglected in current discussions.

The problem of public broadcasting is very directly connected to the defective political structure of western democracies. For the political system of indirect representative democracy has often led to the emergence of a political apparatus in which the elected powers are in fact increasingly less representative. As far as Belgium is concerned, this lack of representativeness is, moreover, linked to a far-reaching segmentation, not only in the political, but in the linguistic sphere as well. All this means that public broadcasting is controlled by political parties, which are not capable of achieving a truly 'public' role, but which in the main consider the public broadcasting service as an instrument for the benefit of their own party-political strategy. The breeding-ground for the crisis of public broadcasting is therefore largely to be found in the crisis of the representativeness of political power.

Notwithstanding the diminishing representativeness of political power, parties and governments have in the past always striven continually to extend their grip on the audiovisual media. However, the increasing political control of the audiovisual media should not be explained in a subjective, personalised way. it should rather be connected with the first premise of this essay, namely the increasing expansion of exchange value. For this also has a political dimension: the expansion of exchange value has led to a social system in which the power to make decisions regarding fundamental issues of social policy is increasingly being taken away from the political organisms created to that end, and has come to rest with the highest offices of multinationals and financial concerns. Thus, political power has clearly lost in influence to the benefit of private economic power. This shift in decision-making power has important implications for the activities of politicians. Since politicians can only insufficiently work in a policy-determining way, they, out of political self-interest, particularly direct their attention to 'image building' in relation to that policy. Since they may politically be called to account for a policy of which they can barely determine the central themes themselves, they will as a result make a particular effort to gain control of those institutions which channel that policy to the audience at large. The high degree of politicisation of the public broadcasting service, seen in Belgium as one of the main causes of the crisis of public broadcasting, is largely due to the hollowing out of political power by the increasing expansion of exchange value.

The concept of 'public broadcasting', as it has grown historically, is part of 'welfare state' ideology, providing a social service for all on a non-profit and pluralistic basis. As such, the public broadcasting service is an ideological construction that in the first place had a legitimising function. On this level a comparison can be made with the social security system, which is also a constructive element of welfare state ideology and which is, at this moment, also coming under fierce attack.

The crisis of public broadcasting therefore has deeper-lying reasons than competition from the new media technology. This competition has only had an accelerating effect in the bringing to the surface of a 'crisis' whose causes were present long beforehand. In the current political debate on these matters, these structural causes are hardly touched upon. The strategy contrived by policy-makers to save public broadcasting also clearly suffers from the same symptoms.

The Crisis of Public Broadcasting: The Strategy
The strategy discussion concerning the position of public broadcasting on the policy level is in the main limited to the problem of advertising. The Board of Directors of the Flemish public broadcasting service (BRT), which till now has refrained from broadcasting advertisements, is now trying to capitalise on advertising revenues in order to improve its competitive position in respect of the commercial station to be created. The Socialist Party, too, which was always dead

set against introducing advertising on the public broadcasting service, has completely reversed its attitude since 1981. Since advertising on television is inevitable, it is better to allow this to the BRT, which will then reap the financial benefits – such is the train of thought of a large part of the left. This proposition does indeed seem most appealing. It is precisely therefore all the more important to point out the dangers involved in such a standpoint. Both on the level of strategy (in which direction does one want to go with public broadcasting in the long run?) and of tactics (how can the financial problems of the public broadcasting service be helped in the short term?), the disadvantages of introducing advertising on the public broadcasting service could well be more decisive.

As far as long-term strategy is concerned, it is clear that any introduction of advertising on a television station has irrevocable and far-reaching effects on the programming policy of that station. The provision of pluralistic programmes, which is also regarded as a basic principle for a public broadcasting service, is extremely hampered under these circumstances. This argument against advertising is obviously very old but has as yet not been disproved. On the contrary, the experiences of the last few years in Italy, Canada and Great Britain once more confirm this.

The pro-advertising standpoint, however, seems rather to be prompted by tactical considerations. The attraction of advertising monies must then be seen as a pragmatic solution to help the BRT out of its financial problems, even though one is aware of the possible threat to the principle of public broadcasting itself. But even on this mere pragmatic-tactical level it is very much in question whether advertising monies as a supplementary revenue source for public broadcasting will prove to be such a success. For those same factors that threaten the livelihood of a Flemish commercial station also apply to the BRT: the advertising market in Flanders is very limited (at the most BF 2.5 billion advertising revenue yearly), an important part will have to be handed over to the press, and in the near future commercial satellite television will appropriate a large part of advertising revenue. Given these circumstances, not much of the advertising budget will be left for the BRT, especially not to build up a popular service of its own.

This is all the more poignant in that the same BF 2.5 billion extra income can also be obtained in another, safer way. For at present BRT receives about 55% of the radio and television licence fee revenue. If this percentage were increased to 80%, the public broadcasting service would immediately gain an added revenue of BF 2.5 billion, which does not have to be shared with the press; is not subject to competition from satellite television; would not, moreover, affect the principles of public broadcasting. However, this is a political matter that must be settled by government. Nevertheless, the public channel could substantially strengthen its negotiating position if it stopped trying to attract advertising monies, and quietly left these for the new commercial channel. As compensation for this a fixed and *high* percentage of licence revenue (minimum 80%) could be claimed on just grounds. However, as soon as the BRT accepts advertising monies, it will at the same time lose every argument for ever receiving even a minimal increase in its share of the radio and television licence fee. What's more, as soon as the first adverts appear on the public broadcasting service, Liberal Party voices will be heard bidding to restrict the licence fee or even to abolish it in future.

So if advertising monies were attributed to the BRT, a scenario threatens to develop that could undermine the position of the public channel. On the one hand large parts of the advertising cake will in the future be eaten up by commercial satellite television, leaving the public channel with the crumbs. There is a real chance on the other hand that the radio and television licence fee will be sharply reduced under Liberal Party pressure. The plea for advertising on the public broadcasting service could well have exactly the opposite effect: instead of leading to the salvation of the BRT, there is a much greater risk that it will lead to a financial drain on the public broadcasting service.

Towards an Offensive and Alternative Media Policy
Within this outline framework of the structural causes of the crisis in public broadcasting and in the headlong evolution towards an information society, the problem of broadcast advertising is obviously but one small component. The attitude of the organised left on this matter is, however, characteristic of its global media vision, which has traditionally too often been distinguished by a merely defensive attitude and a wrongly measured pragmatism. With an eye on future media developments it is most important that policies be drawn up which will tackle the media problem in a fundamental way.

Concerning financing policy, it is of the utmost importance here to counter the prevailing opinion that the future media constellation will either have to be financed by the audience (through licence fees) or by advertising. This is obviously a false choice because in both cases it is the tax-payer or the consumer who foots the bill (the advertising tariffs are naturally passed on in terms of the prices paid by the consumer). The private companies which in future will be directly involved with the building up of the information society are already making handsome profits. The three largest Belgian electricity companies, which have control over a large part of the cable companies (Intercom, Ebes, Unerg), in 1984 registered an aggregate net profit of BF 16.4 billion. These net profits amount to about three

263

times the working budget of the BRT in 1985! An alternative model of financing would gain much credibility if the financial sources were sought in that direction. For the main principle of an alternative finance policy is that the profits which are made by operating the media system (from hardware as well as software) would be used for a balanced upgrading of this media system, and would no longer serve to swell the already well-lined coffers of the financial houses and energy companies. It is clear that the pro-advertising standpoint cannot be part of such an alternative model of financing, but rather anticipates a further commercialisation of the media, such as already envisaged by the Conservative-Liberal majority. Formulating such financing models obviously takes some political courage from politicians, as is also the case regarding other aspects of media policy, for that matter.

For at the same time as outlining another finance policy, the structural-organisational policy regarding public broadcasting must be thoroughly re-thought. An offensive strategy is strongly called for here. Is there any point fighting for the 'defence' of the public broadcasting organisation, if it turns out that a truly public broadcasting service has never existed? The main mission for the future is therefore not so much the defence of the public broadcasting service, but rather the creation of a new service, adapted to the fresh tasks which await public broadcasting in the future media system. In the first place serious endeavours will have to be made for a fundamental de-politicisation and de-bureaucratising of public broadcasting structures.

As long as party-politicians, from the Right as from the Left, consider public broadcasting as an instrument for the benefit of a party-political strategy, the public broadcasting service will remain doomed. It is therefore in the hands of the party-political camp: either the politicians stubbornly cling to their political hold on the Flemish audio-visual constellation, which in this case will in due course be eradicated by international media groups; or they summon up the political courage and maturity to grant the public broadcasting service the necessary financial means and autonomy that will enable it to create an image culture of its own.

Moreover, a fundamental restructuring of the public broadcasting service must reach further than a rewording of aims, control and access. This not only has to do with a review of the administrative and institutional framework; the internal relations of production are also ripe for reconsideration.[16] For broadcasting is indeed not merely to be thought of as an economic system which produces merchandise, but at the same time as a system of representation producing meanings. It is especially important, with a view to formulating an alternative media policy, that the political economists and the culturalists try to meet each other with this 'dual function' of the media clearly in mind.

References

1. See H. Verstraeten, 'Bijdrage voor een politieke economie van de massamedia in Belgie: Inleiding tot de theorie' ['Contribution to the Political Economy of the Mass Media in Belgium: Introduction to the Theory'], in *Massacommunicatie* vol. 7 no. 4, 1979.
2. See H. Verstraeten, 'De nieuwe media en het laatkapitalisme: Een politiek-economisch analysekader' ['The New Media and Late Capitalism: A Political-Economic Framework for Analysis'], in *Vlaams Marxistisch Tijdschrift*, vol. 17 no. 5, 1983.
3. See H. Braverman, *Labor and Monopoly Capital: The Degradation of Work in the Twentieth Century* (New York/London: Monthly Review Press, 1974); P. Edwards, *Contested Terrain: The Transformation of the Workplace in the Twentieth Century* (New York: Basic Books, 1979); D. Ernst, *The Global Race in Microelectronics: Innovation and Corporate Strategies in a Period of Crisis* (Frankfurt-am-Main: Campus Verlag, 1983).
4. M. Tyler, 'Telecommunications and Productivity: The Need and the Opportunity', in M. L. Moss (ed.), *Telecommunications and Productivity* (London: Addison-Wesley, 1981), pp. 5–6.
5. T. Hollins, *Beyond Broadcasting: Into the Cable Age* (London: British Film Institute, 1984), p. 27.
6. See H. Verstraeten, '100 Jaar socialistische pers in Belgie' ['A Century of the Socialist Press in Belgium'], in *Vlaams Marxistisch Tijdschrift*, vol. 19 no. 1, 1965.
7. See J. C. Burgelman, 'Omroep en objectiviteit: De politieke agenda-setting functie van de externe kritiek op de omroep (radio) berichtgeving (periode 1970–1977)' ['Broadcasting and Objectivity: The Political Agenda-Setting Function of External Criticism of Radio News, 1970–1977'], in *Res Publica*, vol. 26 no. 2, 1984.
8. C. Dardenne, *La presse et une télévision flamande commerciale* (Brussels, 1983).
9. For a broad analysis of the economic structure of RTL, see E. Lentzen, 'Compagnie luxembourgeoise de télédiffusion "CLT"', in *Courrier Hebdomadaire du CRISP*, no. 1066, 18 January 1985.
10. See Belgische Vereniging van Dagbladuitgevers/Nationale Federatie der Informatie Weekbladen, *Onderzoek naar de gevolgen van de invoering van TV-reklame voor de geschreven pers* (Brussels, 1981).
11. See H. Verstraeten, 'Pers en commerciele televisie: medespelers of tegenspelers op de advertentiemarkt?' ['The Press and Commercial Television: Partners or Opponents in the Advertising Market?'], in K. Raes (ed.), *De nieuwe media: culturele politiek en politieke economie van de nieuwe media* [*The New Media: cultural Policy and Political Economy of the New Media*] (Ghent: Masereelfonds, 1985).
12. More information concerning the economic restructuring of the Belgian holdings can be found in 'La concentration économique et les groupes Société Générale de Belgique, Cobepa, Bruxelles Lambert et Frère-Bourgeois en 1981–1982', in *Courrier Hebdomadaire du CRISP*, nos. 993–4, 1983.
13. See A. Lange, 'Skizze eines ersten weltweiten Fernseh-Networks', in *Media Perspektiven*, no. 2, 1986.
14. 'OTV zet aktiviteiten op een waakvlam', *De Standaard*, 3 March 1986.

15. See Hollins, *Beyond Broadcasting*, pp. 35–50, for interesting information on this point.
16. C. Gardner, J. Sheppard, 'Transforming Television: Part 1: The Limits of Left Policy', *Screen*, vol. 25 no. 2, 1984.

WILLIAM H. MELODY

Pan European Television: Commercial and Cultural Implications of European Satellites

As a result of continuing improvements in satellite and related communication technologies, international television is taking on greater significance than ever.[1] It is focusing our attention, perhaps for
the first time, on the issue of what truly 'international' television is,
and what its implications might be. Most research in the past has
viewed international television as the exchange among countries of
national programmes produced for domestic consumption. International programming could be catalogued by country of origin. Many
research studies have documented the national diversity of international programmes, or the dominance of this international market in
national programmes by US suppliers and programme producers.

Upon reflection, however, it is apparent that the international
television programme market has been a market for the exchange of
programmes that were designed essentially to serve some country's
specific domestic market conditions and/or public service objectives.
The programmes were designed neither to meet the requirements of
an international or global market, nor to satisfy the objectives of an
international or global public broadcasting service. Television programming that is specifically designed for the precise purpose of
responding to international or global market conditions, and not to
the particular domestic conditions of any country, must be different –
perhaps substantially different – than the programming produced for
the domestic market in any particular country.

Radio broadcasting provides an illustration of the considerable difference. The programming for the BBC World Service is substantially different from the programming for the BBC domestic channels.
With radio broadcasting, both the international and the domestic
markets can be served with programming specifically designed for
each market. With television, however, the very high costs of programme production make it extremely doubtful that separate programming for both markets can be maintained. The vast majority of
programmes will be produced for the dominant market, either
domestic or international, and sold in both the dominant and subsidiary markets.

Television programmes designed at the outset for global marketing and viewing are now in the process of being developed. They are made possible by the implementation of a variety of new communication technologies, the most significant of which are satellites and broadband cable (CATV).[2] Programme development increasingly is being financed by joint ventures of production houses in different countries. This facilitates the development of international programming that will take into consideration different viewer tastes and habits in the variety of countries in which the programming will be viewed. It also smoothes the difficult path of access to television markets in those countries. This paper explores some of the potential implications of the evolving international television market, with particular reference to the implications of pan-European television broadcasting for the promotion of a European common market and a European common culture and consciousness.

The Institutional Constraints on Programme Content
The characteristics of media content are heavily influenced by the structure of the institutions that make up the total broadcast system. We are generally familiar with the major differences in programming content produced by commercial and public service broadcasting. But there is great variation within both the commercial and public service models. Commercial broadcasting is not totally ruled by profit maximisation and often provides some public service programming. Historically there have been limits on the extent to which commercialisation is permitted to penetrate programme content. For example, there is normally an observable separation between the programmes and the advertisements.

Similarly, public service objectives are defined differently in different countries, and are constrained in varying degrees by cost and audience criteria. Some public service broadcast organisations emphasise a national public service while others emphasise a regional or local public service. Among those emphasising a national public service, the BBC, in the Reithian tradition, historically has interpreted its public service mandate as a paternalistic educational uplifting of the masses, a view that has been under assessment for some time within and without the BBC.[3] The Canadian Broadcasting Corporation (CBC) mandate is to promote national unity, to develop a national consciousness and to interpret Canada for Canadians.[4] The US Public Broadcasting Service (PBS) objective is primarily to promote artistic, cultural, public affairs and related programming as a supplement to the US commercial broadcast system. Other countries have different models of public service broadcasting.

In addition, the system of finance, the structural relations with production houses, the standards of accountability employed and other factors all affect the programme content. These institutional constraints do not deny creativity and discretion so much as channel it in certain directions. Some of the most creative programming is channelled into advertisements.

The development of an international or global television market is new. It opens opportunities, but the pursuit of these opportunities will have significant implications for all television programming. There has been very little serious examination as to how global television markets are likely to develop, what the characteristics of international or global television are likely to be, and what the economic, social, political and cultural implications might be. This is an extremely important issue because it is possible that programming for global markets will displace many traditional forms of programming for domestic markets, including both commercial and public service programming. The BBC, for example, has a potentially lucrative opportunity to enter global television markets, substituting programmes that have been produced for the global commercial market for a major portion of its traditional domestic public service programming that it has developed historically for its United Kingdom audience.

The CBC is being forced into international programming. For many years the CBC has imported the great majority of its prime time entertainment television programmes from the US because of the high cost of programme production, the relatively low cost of purchasing US programmes and ever-tightening budget constraints. The CBC has been instructed by the Canadian government to direct its programming to global markets, and therefore by implication, away from unique Canadian conditions, as a means of increasing revenue and thereby reducing its subsidy from the public treasury.

Even in the US, despite its enormous domestic market, producers are no longer simply trying to sell in foreign markets television programmes that were produced for US domestic consumption. They too are moving to a condition where they are attempting to produce programmes to meet the criteria of a global market that they also can sell in the US domestic market. High volume programme-producers in many countries – including most major public service institutions – are now developing their strategies to exploit the 'export' potential of television programmes as part of national industrial policies.

As with all cases of institutional change, there will be both positive and negative implications, no matter what one's point of view or ideological stance.[5] For example, if a European country fears an invasion of US commercial programmes, it may be relieved (slightly) to learn that the programmes coming from US producers may not be carrying US domestic commercial and cultural values, but rather the conception that US producers hold of widespread international or global commercial and cultural values. But if one seeks programme diversity within one's own country that is responsive to a variety of specific domestic conditions and issues, there is likely to be disappointment upon learning that the incentives of national public service programmers may be moving away from that objective, towards the objective of success in global markets. And success in global markets is likely to be judged by commercial, not public service standards.

Will global television programming help overcome nationalist programming differences and lead to programmes designed to unite us all by emphasising cross-national common cultural values and the flowering of diversity? Or will it lead to programming designed to serve an artificial bland, homogeneous set of values that represent no one and offend no one, that is, a destruction of diversity to serve only the lowest common denominator? What types of sub-market programme specialisations are likely to develop, for example, children's, soaps, news, public affairs, etc.? Will they be responding to commercial objectives, public service objectives, or a mixture of both? What will be the extent of commercial penetration of the international television market? Will there be international public service programmes? How significant is the displacement of domestic programming likely to be in different countries, and what will be the implications for particular types of traditional domestic programming? These are only some of the questions that remain largely unexplored.

Changing Characteristics of Public Service Broadcasting
All countries initially established radio and television broadcasting with a strong commitment to public service. Even in the US commercial broadcast system, a variety of public service obligations traditionally have been imposed upon the broadcasters by the regulatory authority, the Federal Communications Commission (FCC). In almost all countries the financing of public service programming is integrally tied to the domestic audience, that is, the 'public' that is being served and that is providing the financial support. The import of large volumes of foreign commercial programming into most countries must affect traditional public service broadcasting in a negative manner, simply because it provides an alternative that inevitably must reduce the audience for the public service programmes.

This raises an important issue of public policy. Do global television markets inherently require the decline of national public service broadcasting by a direct substitution effect? If so, is there a case then to be made for the establishment of international public service institutions to supply European or even global public service programming? As national public service institutions, are the BBC, the CBC, or other national public broadcast organisations remaining true to the spirit of their public service obligations if they now begin to develop their 'public service' programming primarily for global commercial television markets rather than to meet the objectives of either a domestic or a global public service?

With the introduction of pan-European commercial television by satellite, why have the dominant national public service broadcast institutions in Europe not responded with an aggressively supported European public service alternative? Is this not where one would look for the potential positive cultivation of the desired common European cultural values that are now so often discussed as abstract theoretical notions?

Historically, the geographic limits of broadcast markets have been extended from local to regional, national, international, multinational and supranational by a series of remarkable improvements in communication technologies. 1986 was the 60th anniversary of Logie Baird's first demonstration of his 'television' electronic communication signal. But the age of electronic communication really began on 24 May 1844 when Samuel Morse sent the first telegraph message, 'What hath God wrought!'. Morse undoubtedly would be astounded to see the technological improvements in electronic transmission capability that have occurred over the past century and a half. And indeed many writers seem to attribute these technological developments to a 'God' of some kind, inviting individuals, and entire societies to submit themselves to technological determinism.

Is public service broadcasting to be rendered obsolete by the march of modern communication technologies? Or are the technologies simply the tools that provide opportunities to change the traditional institutional structure? Perhaps the technologies are providing opportunities that commercial television entrepreneurs are exploiting fully, while public service institutions are being left at the starting gate. Maybe our existing public service institutions are having difficulty adapting effectively and defining a clear role for public service television in the new institutional environment. This is an important issue worthy of more detailed examination.

Technological Determinism or Institutional Rigidity?

It is noteworthy that God never won the Nobel Prize, and that the inexorable march of technology is very much a 'forced march'. It is directed by powerful institutional interests seeking to improve their condition – generally at the expense of other powerful institutional interests. The new technologies are tools that have provided particular institutions with the leverage to force major alterations in economic, political and social power relations to the advantage of those institutions applying the technologies. But clearly new technologies

271

are not the only factors that facilitate or retard change in institutional structures. And the direction of technological application is very much influenced by other institutional factors.

History is strewn with illustrations both of technological failures and of demonstrably useful technological improvements that have never been introduced. For an illustration we need look no further than a recent technology that has affected most of us, the personal computer (PC). The arrangement of the keys on the English language PC keyboard is inherited from the typewriter. As many of you may have observed at certain moments of frustration, the keys do not seem to be optimally arranged for efficient use. They aren't! Detailed studies of the frequency of use of letters, letter combinations, manual dexterity and other factors affecting the speed, efficiency and comfort of typing have provided a basis for an improved arrangement of the keyboard for many many years. But the chances of adoption remain slim. Adoption of a new keyboard arrangement would upset many powerful vested interests, make obsolete enormous amounts of investment, de-skill millions of people, impose substantial adjustment costs and only succeed if all the major established firms in the field adopted it simultaneously. The economic, social and cultural barriers to date have been sufficient to prevent change. In the study of new technologies, an area of research taking on increasing significance is the particular types of institutional arrangements that promote, retard, direct or misdirect the development and application of potential new technologies.

Although Western societies have so institutionalised modern communication technologies that they are becoming ever more dependent on them, and most studies of technology tend to treat it as an exogenous force over which one can exercise little influence, there are in fact wide areas of discretion. These are reflected primarily in policy choices that significantly influence: the direction and speed of implementation of the new technologies; the institutional arrangements for managing and controlling them; the type of product and service applications that are developed; and the characteristics of the information content that is produced and transmitted using those technologies. What may appear to be cataclysmic upheavals caused by a conjunction of major new technologies may create great threats, but they also create great opportunities. This always has been the case. But for the future the stakes for society are much higher than they have been in the past. Therefore, research directed to revealing the full range of policy options and their long-term implications for society is essential.

This is crucially important for the noncommercial aspects of technological and institutional change. The pace and direction of technological change is often, but not always, governed primarily by commercial considerations. Unless noncommercial interests recognise the significance of the potential changes, and seize the opportunity to turn the application of new technologies to serve their objectives, the ground may be swallowed up by commercial applications. Clearly when radio and television broadcasting were first developed, their potential as tools for education, public service and influencing public opinion was recognised and acted upon by governments in many countries before the commercial potential was perceived and acted upon.

This is not an argument for the artificial restriction of commercial applications of new technologies. It is an argument for noncommercial, and especially public service institutions, to take a much more serious and active examination of the opportunities created. They too must adapt their traditional roles to the new environment so that their objectives can be met effectively, and perhaps in innovative ways.

Pan-European broadcasting requires public service broadcasting to redefine its role under the new conditions of satellites (including direct broadcast satellites), broadband cable and VCRs. It requires public service broadcasting institutions, like the BBC, to redefine their roles in light of the new continental and global broadcasting opportunities. It requires innovative independent research on the options and implications of different potential paths of development; and it very likely requires different policies for guiding that development.

The Breakdown of Discipline-Based Research
The issues generated by these technological and institutional changes cut across many research professions. This is unquestionably one major reason why such a diverse array of research interests and professions were assembled at such a timely and notable International Television Studies Conference. Researchers from a wide variety of backgrounds and interests have been thrust together by a common recognition that the formerly stable ground upon which they have been doing their research is shifting. The most cherished fundamental assumptions of our generally accepted theoretical models have been rendered irrelevant. Our solid empirical findings have rapidly been made obsolete. The anchor that ties us to our traditional professional disciplines has been cut and we are at sea. But is this threat or opportunity?

The professional disciplinary classifications always have been artificial creations, especially in the human and social sciences. The boundaries between professions are arbitrary, and they shift under the pressure of changes in the structure of research and teaching institutions. When the major institutions in society are relatively stable, professional specialisation often is productive because it directs research and analysis to an intense examination of a narrow range of issues. And this has generated much valuable knowledge. But when institutions are undergoing change, specialisation often is transformed into myopia. As our disciplinary ships head into the storm, we can doggedly maintain our focus on the issues historically determined for us by the disciplinary boundaries established under stable institutional arrangements that are rapidly becoming obsolete. But this may restrict the range of issues that we are capable of addressing to the relative significance and duration of a plan for developing the optimal arrangement of the deck chairs on our rolling ship. We are likely to be in a position to make a better contribution if we pay more serious attention to the dynamic changes occurring in the system, such as the force of the waves, the nature of the changing weather pattern, and the navigation possibilities.

Perhaps fortunately, the study of communication always has cut across traditional disciplines, although most researchers have studied it from the perspective of a particular discipline. The new technological developments emphasise the need for research that cuts across the established disciplines. In parallel to the developments in broadcasting technologies, institutions and markets, research in the field must become more interdisciplinary, multidisciplinary, transdisciplinary and even supradisciplinary. The institutional structure of research must adapt to the changing character of the institutions under study. It must be so if a comprehensive understanding is to be achieved, if policy research is to be relevant and if policy and management decisions are to be informed.

'Television Without Frontiers'

An extremely important part of current developments with respect of global telecommunication systems and broadcast markets involves the establishment of pan-European television broadcasting, either to cable television head-ends or to direct broadcast satellite receivers. This is being actively promoted by the European Economic Community (EEC) as part of its mission to knock down artificial barriers among member countries in order to enhance the creation of a common European market, a common European identity, a common European presence in the world, and the flowering of a European culture that is distinct from a collection of European national cultures.

In June 1984 the EEC Commission published a Green Paper on the establishment of a common market in broadcasting, especially by satellite and cable.[6] Entitled *Television without Frontiers*, it aims to illustrate the relevance of the EEC Treaty to broadcasting, and invites discussion as a basis for possible future proposals for legislation that would attempt to unify the present diversity among national laws and regulations of the member countries concerning broadcasting, copyright and related matters.

The Green Paper is based upon earlier resolutions of the European Parliament. The earlier resolutions were directed toward promoting European unity by creating programmes on European affairs, including a European television channel, that is, a type of European public service. But this idea was resisted by the national broadcasters in the member countries, including both public and private sector broadcasters. The primary stated concern was that it would give politicians undue influence over its content. The unstated concern was the creation of potential competition for the established broadcast organisations, public and private.

As with many EEC initiatives, attempts to get the member countries to yield some national power in the interest of a European focus faltered. Few institutions give up power voluntarily, especially national governments. This has been an historic problem for the EEC. Most national governments see it as a competitor for power and influence in their respective countries. Of course, it is! But there cannot be a common European approach to anything unless national governments and national institutions are willing to yield sufficient power to permit a European approach to develop.

The Green Paper seizes on the satellite and cable technologies as a way of creating a European focus, a European common market in television broadcasting. The new technologies will assist new television services that will be made pan-European by the nature of the distributional technologies. As a new European-wide television service, it need not raise the difficult issues associated with entrenched national institutions and established national television services. Rather it will require only liberalisation of controls in order that national restrictions should not obstruct the operation of the common market.

The major obstruction that concerns the EEC is national control over broadcasting signals that can be received within the country. Nations could retain the power to place restrictions on programme production within their respective countries, but the EEC would like to establish common market reception regulations. In essence, these EEC regulations would be those of the EEC country with the least restrictive regulations, or perhaps even less restrictive than all EEC countries, in order to provide access for pan-European television to the maximum amount of programming from all potential sources. Put simply, the EEC compromise is that the individual nations will determine what television programmes are produced in and sent from the respective countries. The EEC will determine what can be received across all the EEC countries.

Placing restrictions on programme production and origination upon one's own domestic producers would simply place them at a competitive disadvantage in competitive global television markets. Within this global market structure, a powerful incentive is created to eliminate all regulations as a basis for maximising a nation's competitive advantage in programme production and origination.

As an alternative, it would be at least theoretically possible to establish a common European television market based upon the regulations of the EEC country with the tightest regulations. But this would restrict competition to a much narrower range of programmes by excluding some programming (including possibly all commercial programmes) from the market. In addition, it could be claimed that the EEC would be imposing the regulations of the most restrictive country on the programming decisions of other countries, thereby denying some countries the opportunity to receive television programming that their national regulations permitted.

If the objective is to promote pan-European broadcasting specifically directed to common European interests, the more restrictive model may be more appropriate. The EEC proposal of minimal regulation does not really promote pan-European television. It promotes global television. The dominant source of programming will be almost certainly the US, and programme production by European producers and broadcasters will be directed not to common European themes, issues and values, but to those of the competitive, commercial global market. There is nothing uniquely European about it, as a casual observation of current pan-European television quickly demonstrates. Under the model of more restrictive regulation, the programming is much more likely to reflect the claimed objective of promoting television programming with a common European cultural dimension.[8]

It must be recognised, of course, that the EEC position on this issue is not determined by a detached assessment of the probable characteristics of the new types of television programming likely to be forthcoming for European viewers. In comparison to the member states, the EEC is not a powerful body. Its major constituents, the national governments, not only view it with circumspection but also keep it on a very short leash. The encouragement of pan-European satellite broadcasting provides an avenue for the EEC to enhance its powers relative to the member states by establishing a degree of commonality across the community. The fact that this commonality in broadcasting may not be a distinct European commonality, but rather a global commonality, is not a significant issue. Rather, it is a policy refinement with which the EEC is not powerful enough to cope at this stage of its development. The EEC is desperately seeking any common ground that it can find as a basis for justifying its legitimacy. If its proposed television policies actually may make it more difficult to promote a distinct common European television culture, so be it. That problem will have to be dealt with later.

A much more important issue for the EEC is the economic viability of pan-European satellites. This is essential to the creation and maintenance of EEC leverage with the European nation states on this issue, and overall. The financial reports indicate that the existing pan-European satellite broadcasters have not done well in the early going. None of them are profitable. Sky Channel, the oldest satellite channel, has lost money continuously since its inception in 1982. There is little, if anything, distinctly European about the television programming distributed by pan-European satellites. But if pan-European satellite television is going to serve the EEC's vested institutional interest, it must survive and become financially viable. To increase the possibility of financial viability, the different television broadcast regulations of the member states must be harmonised at a level that will provide the minimum constraints on achieving viability.

By taking the position that it has, the EEC suddenly has found some powerful new constituents in the form of transnational corporations (TNCs) from several powerful industries, including satellite equipment, cable television, DBS receivers, programme production, global advertising agencies and the largest TNC advertisers who can benefit substantially by being able to advertise in European and global markets. With this kind of powerful corporate support from a new EEC constituency, the EEC is able to transform itself into a real contender for increased power in the political economy of Europe. It has acquired the support of the transnational corporate community.

Financial Viability and Public Service
The regulations of greatest relevance to the financial viability of pan-European satellite television relate of course to advertising. That is the key to financial viability. The fewer the restrictions on advertising, the greater the chances for financial viability. Similarly, any development that would reduce the audience ratings for pan-European commercial television would tend to reduce the financial viability below its potential. Thus, the promotion of a European public service television network would tend to conflict with the objective of promoting the financial viability of pan-European commercial television. In fact, the opportunities of the latter will be enhanced if competitive television at the national level, public service or commercial, is reduced or restricted.

But there has been no mandate for the EEC, or the member states, to sacrifice the role of public service television, or for that matter, to reduce the public service obligations imposed on commercial televi-

277

sion, in order to enhance the financial prospects of pan-European commercial televison. Like many government and public service agencies, the EEC finds itself in a severe contradiction between its broad European public interest objectives, and the policies that it must adopt if it is to enhance its own power and influence in the foreseeable future. If the EEC does not promote the 'television without frontiers' policy, it may not have enough power to establish any common European policy relating to television. Yet, if it does, the EEC may enhance its power, but become dependent upon the support of transnational corporations, and therefore be incapable of achieving its public interest objectives. Under these circumstances, it is unrealistic to assume that the EEC can promote the interests of a common pan-European public service television network effectively.

Are not the national public service television broadcasters in a powerful position to promote pan-European public service television effectively to their national governments, the EEC and other influential parties? Are they not in a powerful position to produce, or commission, European public service television programming? They are. But they have not taken up the cause seriously, it seems, for several fundamental reasons. First, they have the same mistrust as their national governments of anything European that would require them to yield some of their monopoly power in their home countries, in this case over public service broadcasting. Second, the new European public service programmes would be competing most directly with domestic public service programmes in their home markets. Third, the cost of buying in programmes from the global programme market is much less costly than producing new European public service programmes, which under current budget constraints would divert funds from domestic public service programming. Fourth, the enormous overcapacity in global telecommunication transmission and delivery systems that have been, and are being installed, is creating a relative shortage of television programmes on the world market. The BBC in particular is strategically placed to compete profitably in the rapidly developing global television market as a programme seller, because of its enormous inventory of programmes and the high level of its annual production.

This raises a question as to whether the BBC is about to be transformed from a national public service television broadcaster into a UK-hosted and subsidised transnational television company competing in a global commercial television market. If so, identical programmes produced by the BBC will be sold internationally as commercial television and broadcast domestically as public service television. It is now commonplace for the BBC to purchase US commercial programmes for showing on the BBC public service. By not showing advertisements, it is believed that at least most of the commercialisation is removed. But to date, the BBC has not been producing programmes with the prime objective of commercial sales in foreign markets. If this step is taken, it may become less and less possible to distinguish between public service and commercial television. Programme diversity would be diminished.

Finally, we might ask what the dominant characteristics of global television might be if the present course of events unfolds without interference or policy guidance. There are strong indications that the US commercial television model is likely to be extended to the global television market. The major forces behind this likely development include the enormous costs of global television, the increased intensity of international competition and the substantial value of global television for the marketing of products and services of the largest transnational corporations. For public service broadcasting, the increasing absolute costs of global, or pan-European television, and the increasing opportunity cost (that is, the potential profit to be made from an alternative commercial application), clearly threatens its future, at least as it has been known traditionally. The public service broadcasters may be both pressured and seduced into commercial broadcast behaviour, a privatisation of purpose, if not ownership.

In addition, in US commercial television, the distinction between programming and advertising is becoming increasingly blurred. The US regulatory authority, the Federal Communications Commission (FCC), essentially has abandoned any attempt, explicit or implicit, to maintain a separation between programmes and advertisements. In recent years the predesigned penetration of implicit advertising into the programme content has increased dramatically, pioneered of course in children's television.[8] In 1984, a new US commercial children's programme, *Thundercats*, was probably the first programme in which virtually every decision, ranging from the creation of the characters, the selection of programme themes and plots, the actions and behaviour of the characters, their clothes and props, all were designed to merchandise products that would be associated with the programme. With each passing year, the content of prime time entertainment programming, such as *Miami Vice* for example, becomes more heavily penetrated by commercial considerations. Ironically, public service broadcasters cannot accept payment for a programme such as *Thundercats* as an advertisement. But they can purchase it as a programme. Indeed, the BBC has done so.

Conclusion
The increasing pressure resulting from the underlying economic characteristics of global and pan-European television is making it more expensive and difficult to deviate from a narrow commercial path of development. If pan-European television is little more than a

giant 'billboard' for the mass marketing of products of the largest transnational corporations, it raises an important question of policy as to whether the enormous allocation of public resources is justified. Clearly this potential development would not reflect a conscious public policy choice.

The issue is not so much one of private commercial interests insidiously undermining the public interest, as it is public service broadcasters losing sight of their mandate, and government policy-makers and regulators failing to implement the noble principles that are so often enshrined in words, but missing in action.

The creation of public service television, and other alternative forms of broadcasting, all come about in the first instance because of the pressure from educational, cultural and citizen's groups upon government to fashion broadcast policy that is responsive to the variety of interests in society with a diverse selection of programming.[9] Pan-European satellites are contributing to a fundamental alteration of historic institutional arrangements in the broadcasting industries. It is time once again for educational, public service citizen's groups: to ensure that their interests are not ignored in the transition to a new institutional structure in broadcasting; to raise again the issue of the purpose, scope and limitations of public service television with the public service broadcasters; to provide closer public accountability for the performance both of public service broadcasters and of national and EEC policy-makers in implementing existing public service and public interest objectives; and to begin to examine rigorously policy options for the effective implementation of public service television in the new institutional structure. All of these tasks require informed analysis based upon in-depth research.

References

1. See, for example, A. Mattelart, X. Delcourt and M. Mattelart, *International Image Markets* (London: Comedia, 1984); G. Wedell, 'The End of Media Nationalism in Europe', *Intermedia*, vol. 11, no. 4/5, Summer 1983; W. H. Melody, 'Direct Broadcast Satellites: The Canadian Experience', in *Satellite Communication: National Media Systems and International Communication Policy* (Hamburg: Hans Bredow Institute, 1983).
2. M. Ferguson (ed.), *New Communication Technologies and the Public Interest* (London: Sage Publications, 1985).
3. See A. Briggs, *History of Broadcasting in the United Kingdom, Volume I: The Birth of Broadcasting* (Oxford University Press, 1961), *Volume II: The Golden Age of Wireless* (Oxford University Press, 1965); G. Wedell, *Broadcasting and Public Policy* (London: Michael Joseph, 1968); Broadcasting Research Unit, *The Public Service Ideal in British Broadcasting: Main Principles* (London: BRU, 1986); and C. MacCabe and O. Stewart (eds.), *The BBC and Public Service Broadcasting* (Manchester University Press, 1986).
4. W. H. Melody, 'The Canadian Broadcasting Corporation's Contribution to Canadian Culture', in *The Royal Society of Arts Journal*, vol. 135, no. 5368,

March 1987; and *Report of the Task Force on Broadcasting Policy* (Caplan-Sauvageau Task Force), Ottawa, 1986.

5. P. Drummond and R. Paterson (eds.), *Television in Transition* (London: BFI Publishing, 1986).

6. Commission of the European Communities, *Television without Frontiers* (Commission of the European Communities, 1984).

7. For an explanation of some aspects of this issue, see G. Murdoch, 'Cultural Policy and Consumer Choice in the Inside "New" Television Age: From Rhetoric to Realities', in *Technological Development and Cultural Policy* (Council of Europe, Cultural Policy Series no. 5, 1984).

8. W. H. Melody, *Children's Television: The Economics of Exploitation* (New Haven: Yale University Press, 1973), and W. H. Melody and W. Ehrlich, 'Children's TV Commercials: The Vanishing Policy Options', *Journal of Communication*, vol. 24, 1974.

9. N. Garnham, 'The Media and the Public Sphere', in P. Golding, G. Murdoch and P. Schlesinger (eds.), *Communicating Politics: Mass Communication and the Political Process* (University of Leicester Press, 1986).

VIII   IDENTITIES AND THE POLITICAL: AFRICA AND SOUTH-EAST ASIA

KYALO MATIVO

The Role of Communication in Alternative Development Strategies

The economic value of mass media is now an established fact. The reason for this is simple: mass media constitute the means by which economic ideas are propagated and justified. It is for the same reason that, in the post-independence period, Africa has become a test site for a medley of theories of development following wave after wave of invasions by countless numbers of western academic researchers. They come in all shapes and colours, brandishing ultra-modern information-gathering gadgets. Their motives are not always known, neither have their harmful effects been properly assessed. Some of them are known to have abused every single privilege granted them by 'host' countries. One unamusing story tells of a foreign researcher who raided a whole national library, emptying it of all the historical documents. The abuses notwithstanding, these alien intellectuals have amassed a solid wealth of information about African peoples, their customs, social systems, languages, history, geography, natural resources, etc. The point cannot be exaggerated, therefore, that information possesses a definite material value, and that the real beneficiaries of the material information gathered in Africa have not been Africans. The question is irrelevant whether or not the information collectors are conscious of their role in the perpetuation of what has been described as a neo-colonial set-up in Africa.

This article examines the part played by communication as an economic factor. The analysis addresses three main issues. The first constitutes perceptions and definitions of the subject-matter. In this connection, the early concepts of development provide the basis for a critical analysis of the assumptions underlying the corresponding models of social development. The second set of considerations relates to the factors leading to the recent attempt at alternative models of development. The task is to discover, within those models themselves, the theoretical concepts whose premises the reality of neo-colonialism has now proved false. The third main section of this analysis deals with the current search for new formulations of communication theory. Following the collapse of the Gross National Product (GNP) as the economic framework of earlier communication theories, new formulations of communication and development theories are now in the making. But contrary to earlier perceptions which tended to treat the two disciplines separately – with an emphasis on development theory – the trend is now the other way. Starting with communication theory, the aim is to arrive at an economic corollary.

Our general framework is as follows:

– as an economic factor, communication is subject to manipulation to suit special interests;
– indiscriminate application of the so-called 'advanced' technology for the purposes of development in the Third World can result in, and has indeed produced, the very opposite of that intention;
– the ownership of the means of communication corresponds to the ownership of the means of production. Hence an independent communication system is a measure of political and economic independence;
– the current struggle by most developing countries for a new information and economic order is therefore in effect a political struggle;
– for the Third World in general and for Africa in particular, traditional forms of communication systems constitute the basis for the reformulation of an appropriate communication theory;
– the rural African societies are endowed with systems of communication which can be put to good use to chart an indigenous path of social development;
– the failure of the developmentalist strategies designed in the West for Africa could actually turn out to be a blessing in disguise. With their backs to the wall, the African people are now forced to explore alternative models, and in the process perhaps establish and consolidate an indigenous framework for social development. But that eventuality presupposes political resolutions, both at the local and international level. The new economic order becomes the bedrock for the effort.

We can now address the subject proper.

The Ghanaian philosopher Kwasi Wiredu defines communications as 'the transference of thought-content from person or group of persons to another person or group of persons'. And 'thought-content' itself is described as 'a semantic unit which invites appraisal in terms of truth or falsity'.[1] Language comes in here as the means by which this

284

'thought-content' is conveyed; and language is in turn mediated through words, gestures, artefacts, signs, sounds, etc. Wiredu takes the opportunity to disparage the concept of objective reality which defines communication in terms of the interpretation attached to 'thought-content'. According to Wiredu, communication is a communal integer, so to speak, without which no human society or community is possible. This community is composed of individuals or interacting persons, and an interaction of persons is only possible on the basis of shared meanings.

Wiredu's concept of communication is faulty. He sees 'truth' or falsity in absolute terms. He fails to account for the process by which 'truth' or 'falsity' are to be established, that is, interpretation. The framework within which an interpretation is carried out and the means by which meaning is to be sought constitute the social point of reference. It is in respect to this reference that a philosophical perception of communication has to be seen to follow one of the main philosophical traditions. Apart from these misconceptions of philosophical perspectives it must be admitted that Wiredu's perception of communication as a process by which 'thought-content' is disseminated is quite correct, if by 'thought-content' he means information or message.

Another attempted definition of communication comes from D. B. Patel. Communication, according to Patel, is simply 'the transmission of messages' following a strict code of conduct constituting five basic elements: the communicator, the message, the medium, the audience/receiver and the impact.[2] But Patel's 'definition' is only a variant of the so-called 'Gerbner model' which attempts to construct a comprehensive theory of communication involving studies of media, measuring the effect of messages, publicity, style and analyses of statements, among other considerations.[3] In both Patel's 'definition' and in the Gerbner model, the 'communicator', the 'medium' and the 'audience' are not defined.

Hans-Peter Kraft offers yet another definition of communication: 'it denotes the spatio-temporal (*raumzeitlich*) transmission of complexities of human knowledge in the form of information between different communication subjects', whereby 'communication subjects' designate the communicators and recipients.[4] By assigning separate but closely linked roles for the communicator, the statement, the medium and the recipient, this concept of communication presupposes a social structure based on socially defined activities. When this concept is then transferred to the wider context of developed-underdeveloped socio-political relations, it becomes clear that information *can* be manufactured just like any other commodity and sold to willing (and unwilling) buyers. And since the industrialised countries possess the infrastructure for the manufacture of information, it stands to reason that they will endeavour to sell this commodity along with others to the non-industrialised countries. The roles are thus internationally defined: the 'communicators' are the industrialised countries, while the underdeveloped world becomes the 'recipient'.

We are now no longer concerned about the mere operations of the communication process, since these can be assumed in any communication system. Our attention is drawn to a complexity of issues emanating from the production and selling of commodities as an economic activity. That information is one of the products of that activity is a function of the mechanism of the entire system. In this context it becomes necessary to define information in its own right. To this end, Kraft ventures to draw a scheme involving a three-step process, namely Signal-News-Information, in which information is understood as 'not only the formal signal, but also the news attached to the signal'.[5] What Kraft does not identify is the value content of information. By 'value content' is meant attitudes, views, emphases and other subjective elements whose validity is limited by the special interests of the 'communicator'.

The inadequacy of these definitions of communication is obvious: the socio-political setting is lacking. But Melvin L. de Fleur and Sandra Ball-Rokeach move in quickly to make amends.[6] From now on the approach shall be one of extracting communication theories from definite ideological frameworks ('paradigms'). These 'paradigms' can be identified as Structural Functionalism, the Evolutionary Perspective and the Social Conflict model. (So-called 'Symbolic Interactionism' and the 'Psychological Framework', which de Fleur and Ball-Rokeach also list as 'paradigms', can easily be treated as subcategories of the main three.)

'Structural Functionalism', or the idea that the structure of the human society tends towards stability, in such a way that 'if disharmony occurs, forces will arise to restore stability',[7] has communication playing the role of the peace-maker between contending forces. About the character of these forces, whence or how they arise, no clues are proffered in the 'paradigm'. The 'Evolutionary Perspective' likens society to a biological organism. The implicit message is that 'social change' is a function of time consequential to interactions and specialisations of society's constituent parts. Social Darwinism is the best expression of this phenomenon. The role of communication in this set-up is to act as the midwife for the safe delivery of every new social formation. But then, if the basic tenets of Social Darwinism are to be respected, 'social change' here coincides with the 'survival of the fittest', and communication is thus incriminated in the matter.

But 'social change' can also occur in the 'Social Conflict model', according to de Fleur and Ball-Rokeach. The 'conflict' character of the model is traditionally blamed on Marx and Engels, but it is to their credit that de Fleur and Ball-Rokeach refuse to be confined to this worn-out cliché, tracing the idea to Hegel, Hobbes and Plato to give it its historical breadth. As de Fleur and Ball-Rokeach understand it, the main assumption in the model is that society consists of social classes whose interests are at variance with one another. The antagonistic character of these classes leads to irreconcilable differences, culminating in conflicts. The ensuing struggle ends in a new social formation. In the midst of these conflicts, communication is found actively taking sides, defending the interests of this or that social class.

Communication theory is thus born as the solution to the problem of the role of communication in society. But this 'solution' is only a 'trivial' one, in the mathematical sense of the word. For 'communication theory' itself needs characterisation. For the purposes of this discussion, 'communication theory' is defined as a depiction of the manner in which a definite people or community produces and disseminates social ideas within a concrete socio-political set-up. The purely normative conceptual infusions which characterise most definitions are consciously divorced from this definition. Here also the futility of striving for a universal theory of communication suggests itself. Even de Fleur and Ball-Rokeach, who necessarily take an American-Euro-centric viewpoint, admit as much.[8] Normative and 'universal' conceptions have a positive part to play in the formulation of any theory; but that part is introduced in those formulations where the aims of the matter under consideration are also universal. Unfortunately that is not the case with communication and development theories.

Like communication, development has also its set of definitions. These are too many and too well-known to reproduce here. Suffice it to offer one which satisfies the needs of this analysis: 'Development is the process through which a society moves to acquire the capacity of enhancing the quality of life of its people, primarily through the solution of its problems.'[9] The importance of this definition lies in its distinction between the acquisition of pure material wealth and the quality of life, so that, as Kleinjans goes on to comment, the affluent countries may in a crucial sense themselves be 'underdeveloped'. It cannot be denied that material wealth is the basis on which the quality of life can be built. But economic development in one country or region can and *has* been achieved at the expense of another. In this case, development requires under-development. The problem can now take on its international dimension.

We are talking about an international economic order whose injustice is so blatant that only very few of its iron-hearted defenders have had the courage to stand up and be counted. The issue has indeed been sufficiently diagnosed and the necessity for change established. We are now far ahead of Walt Rostow's theory of development with its linear five-point stages which were supposed to act as the conveyor belt for the Third World to 'catch up' with the industrialised countries.[10] The misconceptions of that theory have been fully delineated, and we need not repeat them here; suffice it nevertheless to underline its most elementary omission.

Mohammed Bedjaoui criticises Rostow for assuming economic and legal fairness between nations, for ignoring the phenomena of domination and imperialism and for thus reducing underdevelopment 'to a mere question of backwardness which the countries concerned will inevitably make up through the working of laws of world economics'.[11] The uniqueness of underdevelopment, as Bedjaoui rightly argues, is that it is happening for the *first* time in history. Because Rostow fails to understand the nature of underdevelopment, he gives the false impression that the industrialised countries once went through this stage of underdevelopment. The truth, though, is that, prior to the advent of colonialism as a world economic system, there was no precedent of a socio-political nature in which a small group of countries controlled the economic and political lives of a vast majority of others on an international scale. To put it mildly: 'Underdevelopment is really a sort of by-product of the development of the Western countries ... the inevitable result of the development of the system of international economic relations.'[12]

One thing becomes quite obvious: given the present international economic order, underdevelopment is also given. And that indisposition reflects the inability of the developed countries to advance beyond the present stage of their development. Or, as one keen analyst observed:

> Underdevelopment is not merely the product of development, but also, its failure. Development is recent, separate in time and space. Underdevelopment is a product of history and not a natural stage in it – a different phenomenon through which the developed countries have not passed.[13]

It is in this sense that the concept of economic development in terms of absolute economic growth, that is, in per capita GNP increases, turns out to be incomplete. Absolute economic growth implies mass production for mass consumption. This is the practice in the capitalist countries.

Its successful operation relies on the employment of communication as what has now come to be known as economic information. The express purpose of this particular use of communication is to convey information intended to change the thinking and economic behaviour of the people. It is on this basis that indices of economic performance are constructed to measure economic growth rate, the share of capital investments in the national income, volume of energy consumption, employment, etc. But the *social value* of these indices depends on the concepts and notions of economic development assumed. An economic concept which defines itself in terms of the increment of 'goods' and 'services', for example, leaves a yawning gap between the mere availability of material economic products and their social accessibility. As one author candidly puts it:

> Measuring a flow of goods and services is an operation that has specific meaning only when such goods and services are related to the satisfaction of objectively defined human needs, that is to say which can be identified independently of existing social inequalities.[14]

At this juncture we want to look at the various uses of communication for social development. The definition of communication as a means by which information is transmitted from the communicator to the recipient implies that communication serves in such capacities as an educational tool, a medium for promoting culture, a means for dialogue between policy-maker and the people, a way of transmitting messages, a forum for discussing issues on policy and decision-making about development strategies, an instrument for the promotion of community participation, etc. The list can be extended to cover other functions such as 'socialisation', entertainment and so on. But these aspects of social activity are not ends in themselves; they are part and parcel of the whole process of social interaction, and to treat them as separate entities is to deprive them of their content.

We have noted above that the conscious use of communication as an economic catalyst is a particular undertaking of the advanced industrialised countries. There is another form of communication which does not lend itself easily to economic measurements, namely traditional communication systems. The term 'traditional' is used in contradistinction to 'modern' communication systems. The two systems are distinguishable by their communication tools. The 'modern' sector represents the press, radio and television, communication satellites, news agencies, film and video, and theatre. The 'traditional' system can boast of folk dances, songs, drum signals and the market place. In addition, such practices as 'witchcraft' and 'spirit possession' can be shown to function as special forms of communication. It goes without saying that when 'modern' communications are indiscriminately applied in a country where the majority of the people use 'traditional' systems, or at least understand these better, a conflict of sorts develops between the 'modernists' and the 'traditionalists'. Modern communication systems are, by virtue of their origin, commercial in character and represent the interests of their purveyors. In the Third World, where the owners of the means of communication are invariably also the owners of industry – that is to say, largely foreigners – the ideas transmitted through these communication systems are based on non-indigenous social values. Representing, by belonging to, the interests of the 'traditional society', traditional communication systems come into conflict with the needs of this interest group. This conflict materialises as a conflict of values, leading to a clash of definite material interests.

The Iranian case during the Shah's regime provides a classical example of this phenomenon. There, differences of perceptions of development and the attendant strategies for its realisation arose following the Shah's total reliance on Western definitions and values. As a result, a cultural cleavage between Western and traditional values manifested itself in the dramatic events which culminated in the overthrow of the monarchy. The Iranian communications specialist Majid Tehranian described the conflict thus:

> While the monarchical regime defined these [development] goals primarily in terms of quantative leap, in economic growth, the opposition led by radical members of the Islam Ulama called for a return to the fundamental sources of Islamic/Iranian cultural identity – moral legitimacy, political liberty, and social justice.[15]

Here, the conflict between the interests of the Shah and those of his supporters – as defined by Western mass media, on the one hand, and those of the people at large as expressed by traditional communication systems on the other, created a communications irony: for all their sophistication, the modern mass media became uncommunicative, while the 'unsophisticated' traditional systems acquired an infinitely high communication content: 'While the monarchy used mass media as its main channel of *non-interactive* communication, the Ulama used the traditional media of oral, face-to-face and *interactive* communication in the mosques, Hosseiniens and Rowseh-Khanes, etc.'[16] This conflict represented fundamental differences in cultural and philosophical orientation, in which 'politics of identity and legitimacy were thus superimposed upon interest and status politics'.[17]

The element of culture in communication is a constant problem in the study of communication and development. The last word on the subject has perhaps been pronounced by the African revolutionary, Amilcar Cabral:

Culture, like history, is necessarily an expanding and developing phenomenon. Even more important, we must bear in mind that the fundamental characteristics of culture is its close, dependent and reciprocal connexion with the economic and social reality of the environment, with the level of productive forces and the mode of production of the society which created it.[18]

After watching Western programmes on TV, for example, adolescents among the Inuit Indians of Canada started fighting and playing 'tough' among themselves, a form of behaviour hitherto unknown to the Inuit community. For the Inuit and similar cultures, unlike Western societies, the only socially sanctioned 'toughness' relates to human courage in the face of hostile environment, with interpersonal aggression appearing wasteful and foolish.[19] Such modifications of behaviour occasioned by modern communication facilities can result in the disorientation of a people's culture and their social values, thereby depriving them of self-initiative. This is not to deny the possibility of reciprocal enrichment of different cultures. After all, European development took place among a myriad of different cultures, intermingling and modifying each other in the process. As Dieter Senghaas has pointed out, 'Europe's dynamic grew out of a multitude of small social forces competing against each other, long before a century later classical nation states and competitive capitalism emerged, finally subjugating the rest of the world.'[20] Equality of opportunity must be the precondition for the prospect of mutually positive influences between cultures; crisis occurs, as today in the case of many Third World countries, when 'the terms of exposure are so unequal'.[21]

Much of the justification for cultural domination of the Third World by Western social values is based on hackneyed assertions about the 'superiority' of Western technology. It is said that in order to develop, the Third World needs to modernise, and 'modernisation' is synonymous with Western technology.

What is conveniently eschewed in the economic growth scenario is the specification of the exact purpose for which Western technology is to be used. We know that so-called 'high technology' is based on capital and energy-intensive economic concepts. Except, perhaps, for oil-producing countries, most of the underdeveloped countries do not possess the capital requisite for this sort of technology. We know also that what developing countries do possess in immense quantities is untapped human labour. To justify its validity, therefore, one would expect any technology claiming utility for the Third World to be able to absorb this labour power. But as things stand today, much of the technology imposed on developing countries is virtually irrelevant to their development needs.

291

When 'development' in the Third World is taken to mean wester-
nisation of the Third World, the process creates, for its specific purposes, local élites whose political, cultural and social aspirations are
modified to coincide with those of their Western mentors. Their
habits, tastes, ideas, attitudes, life-style – in short their general conduct – sets them apart from the people. For their cultural defence, on
the other hand, the people turn to traditional archetypes, heroes and
martyrs, pitting these against foreign intrusions. The Iranian lesson
cited above seems to have been lost on most Third World countries.
The basic question about the nature and function of 'modern' technology seems to have escaped their attention. But the question may be
difficult to ignore for long since this technology is designed to serve a
definite form of development, that is, the capitalist variety. Unfortunately, contrary to unsubstantiated claims, capitalist development
has not fared very well in the Third World. Problems of poverty,
hunger, illiteracy, disease, coups, etc., are just a few social indices of
capitalist failure in the Third World. The reasons for this failure are
not difficult to find, and a significant number of Western development specialists, who cannot be accused of anti-capitalist sentiments, have admitted as much.

In the first place, capitalist development is a historical phenomenon. Its birth took place under concrete historical conditions,
conditions not only *not* obtaining in the Third World today, but also
irretrievable. Europe was a battle field of *small* but more or less
equal economic forces. Hobsbawm describes the period between 1848
and 1875 as the 'Age of Capital', but actually, by the author's own
admission, the process of capitalisation of Europe had already
started long before that.[22] What is commonly known as the infrastructure for this form of development had already been put in place.
Such was the case with the construction of the railway line, the
development of the chemical industry – for the production of artificial dye-stuffs, the installation of telegraphy, the innovation of steel
production, etc. In other words, the technological basis upon which
the development of capitalism took place in Europe was a consequence of the economic reality of European societies. This technology
was at first a direct product of practical craftsmanship. Until the mid
19th century, most of the technological innovations were made by
people actually engaged in the production process, not laboratory
researchers. Hobsbawm finds this point important enough to emphasise: 'The pioneers of the first industrial phase, Britain and Belgium,
had not been among the most literate of peoples, and their systems of
technological and higher education ... were far from distinguished.'[23] But then, once technology was introduced into the school
curriculum, education became a strategic pilot project for national
development: 'From now on it was to be almost impossible for a

country lacking both mass education and adequate higher educational institutions to become a 'modern economy'.[24] As technology continued to expand to engulf the whole process of capitalist production, the search for new raw materials began, and with it the expansion of capitalism to countries outside Europe, culminating consequently in the age of imperialism. So was also born the Third World. Its role had already been cast and forecast as that of providing the raw material needed to sustain this new economic system.

Secondly, where it has established itself as the dominant economic force in the Third World, capitalist development is a finished product. Further development can only occur along the lines already predicated by the particular requirements of the system, with the Third World squatting on the threshold as 'periphery states'. It is precisely this tight economic framework which hinders any independent capital accumulation in the Third World. It is not true that developing countries do not produce capital: they would be of no use to capitalist countries if they did not. But they cannot accumulate and invest capital independently: 'The assumption that developing countries lack capital is historically refutable: investment facilities were not lacking, they were either squandered or invested abroad owing to lack of profitable prospects in the country in question.'[25] The point is that so-called 'primitive' accumulation cannot take place under the present world economic order for the same reason that capitalism itself cannot start anew in the Third World.

Today capitalism is not concerned with the provision of the necessities of life. Its aim is self-reproduction and expansion. And that means that nothing will be produced under this system unless it can compete profitably in the capitalist world market. There is simply no place for any such thing in the present international economic order.[26]

Thirdly, even if it were possible for the Third World to develop an independent capitalist system of its own, it would need markets for its products. The capitalist system can only survive under the condition of a free market, but today virtually all the markets for products have been taken by the incumbent capitalist order. An independent Third World capitalism would have no chance of breaking into this fortified and heavily defended economic citadel. The impossible mission would require establishing economic beachheads to mount attacks on the existing system. It is superfluous to add that this prospect is unachievable under the present set of world economic relations.

Let us now examine how information comes into the picture as capital. We have stated that the idea of establishing an independent capitalist system in the Third World turns out to be merely an extension of the capitalist system itself. This process is carried out through the use of information as an economic factor. In fact, the term 'information capital' has been coined to refer to such technological implements as typewriters, calculators, computers, telephones, satellites and any other equipment connected with the dissemination of information. This capitalisation of information is a function of the capitalist system as a whole. In its full maturity, the system has now been divided into three main sectors. The 'primary sector' covers agriculture, mining, forestry, fishery, transportation and energy sources. The 'secondary sector' includes industry, manufacturing, finance, insurance, real estate, construction, wholesale and retail. The 'tertiary sector' comprises government services, that is, health, education, entertainment and similar social services. Information capital belongs to the third category, of which it is estimated to make up as much as 70% of the total economic operation.

The importance of the metamorphosis of information into capital has been growing over time in direct proportion to the economic growth in the industrial countries. Thus it is reported that in 1976 information activity accounted for 46% of the GNP in USA. This resulted in a trend of information activity assuming a bigger and bigger role in the US economic development:

> Between 1860 and 1905 the agricultural workers dominated the work force followed by industry, services and information. Between 1905 and 1955, the industry sector took the lead. But by 1955, the information sector became predominant, rising from a low of 15% of the workforce in 1910 to over 40% in 1970.[27]

It is obvious, therefore, that the capitalist economy has been moving from industrial to information economy, a phenomenon clearly reflected in the current 'technological revolution' in computer systems. In the place of an amorphous 'industrial society', a distinct information variety has emerged. But the new scion has a qualitative blemish: it is non-productive. It specialises mainly in administration, entertainment, 'policing' the world, disseminating 'news', etc. It is not by chance then that the 'computer age' coincides with a high rate of unemployment in the capitalist countries. For in a society whose labour force is concentrated in the information services, the ratio of the productive to the non-productive work-force must stand in inverse proportion. This is because the operation of information equipment requires less human labour than, say, mining or the production of steel. The question already suggests itself: in such a society, 'Who would feed, clothe, shelter the society and provide it with cars? Would such an overdeveloped tertiary society not have to turn to an "underdeveloped" primary society elsewhere for the rudimentary essentials of its survival?"[28] It is here in the domain of information economy that alternative ideas about the theory of communication have become necessary. The reasons for this are not difficult to find; ironically, the present economic and information order itself provides most of these. To start with, modern communication systems have had little or no impact on rural populations. This is the case, for instance, with the American Satellite Instrumental Television Experiment (SITE), which was installed in India in 1972. Assessing the effect of this experiment on the Indian rural people, G. N. S. Raghavan notes that because it ignored the cultural values of the people, the experiment could not have had any social meaning for them:

> It is a truth apparent to common sense that decentralised and area-specific programmes, employing the local dialect and depicting the local agro-economic and human landscape, are necessary in any attempt to persuade people to change their attitudes and practices in agriculture or hygiene or, even more so, in family planning.[29]

Another notable failure of a 'high technology' communication system in India was the domestic satellite INSAT I-A. The experiment failed miserably and was abandoned altogether after only one month of existence. Secondly, traditional communication systems, on which any new thinking on communication theory should be based, do not lend themselves easily to manipulation by external forces. This is because these forms are firmly grounded on cultural values which are not easy to dislodge. Furthermore, unlike the modern media, traditional systems are not exposed to external communication forces. What has made the modern media systems in the Third World so important as a means of communication is their total dependence on foreign command centres. For example, even for local matters, the African press tends to rely on Western sources. Graham Mytton recalls a case in the Zambian press in which stories and obituaries on Nkrumah were written by Europeans.[30] And when early this year (1986) Yoweri Museveni took power in Uganda, most people in Africa got their first glimpse of the man through the eyes of foreign news agencies.

The capacity of traditional communication systems to accommodate popular participation in development issues is the third factor rendering these media appropriate as alternative systems to the modern types. Now, the concept of popular participation in development strategies is an old idea.[31] When we say that traditional forms of communication systems have the capacity to accommodate popular participation, we mean that these media possess certain characteristics which are already *normal* forms of social behaviour. The spoken word, for example, occupies a distinctive role in the life of many traditional societies, and in the case of African languages the exchange of proverbs, in particular, can be seen as an important form of 'social communication'.[32] This form of communication is also carried out through songs, legends and myths executed by traditional professionals. In West Africa, for example, griots have traditionally played the role of social communicators. True, they have also performed purely political roles as court jesters and personality cult instrumentalists. But, by virtue of their humble origins, who also possessed the power to make or unmake kings, the griots bridged the communication gap between the rich and the poor. They could, therefore, not afford to be mere mouthpieces of central authorities. Their role as social communicators was only then *socially* valid if the community at large gave them a hearing. If the African griots of yore acted as what Americans would call 'checks and balances' of power, that is certainly not the case in most of the African countries today. In the rare cases where some forms of traditional communication systems are officially recognised and actually given a voice in the modern communication systems – in radio and TV, for example, as in the Ivory Coast – the aim remains to make them serve as means of accumulating power and wealth for the political elites.

Lastly, the indispensable role of indigenous culture in development has already been broached above. We need only emphasise its direct material bearing on the lives of the people.[33] The inability of modern communication systems to have any effect on rural communities, the ability of traditional communication systems to resist external manipulation, their capacity to accommodate popular participation in development activities and their function as permanent cultural barricades against the advance of Western economic ideas: these are only a few examples of the merits of traditional communication systems as alternative points of departure for the formulation of an appropriate communication theory. As indicated earlier, the main fault with existing theories is their futile attempt at universality. The simple fact is that the social set-up determines its communication theory. Even de Fleur and Ball-Rokeach do not disagree with that fact. They are indeed of the opinion that 'Media content must be compatible with the sociocultural context as a whole in order for it to be comprehensive and desirable to an audience large enough for the medium to achieve its economic goals.'[34] Unfortunately (the reasons pertaining to underdevelopment and its concomitant political suppressive machinery) it is not easy to establish an alternative press along the lines of those existing in Europe. In Britain, for example, the Federation of Worker Writers and Community Publishers (FWWCP) has been instrumental in setting up publishing projects independent of the dominant commercial practice. But such ventures are branded 'radical', which in the Third World political parlance translates as 'subversive'. What makes the existence of these press alternatives possible in Europe is, first, the traditional belief in the 'freedom of the press', and secondly the validity of the socio-political establishment. But in the Third World 'freedom of the press' means the denial of it to the vast majority of the people. It means the protection of the local elites against the wrath of the people. It is 'freedom of the press' for the rich and the powerful, a point which Mytton overlooks in his analysis of the subject.[35]

There are of course objective obstacles to the establishment of communication systems in the Third World. Those directly related to this analysis include the lack of research to establish needs; the lack of appropriate training at the University or college level for community media work; political uncertainties about the implications of communal activities necessitated by community media; an inferiority-complex on the part of the educated elite, who expect 'high standards' on a par with commercial media; and the material poverty of the people, resulting in the fact that any activity which does not seem to lead to immediate material benefits tends to be ignored. Yet all evidence points to the traditional systems of communication as the only viable way out of the quagmire in which most Third World countries are now stuck. Examples of their use are almost inexhaustible. Ironic as it may sound, traditional community media can act as a safety valve for the central power base. Following its exclusive reliance on 'modern' media, the Third World has been divided into two nations: the educated elites of the urban areas, and the peasantry of rural areas. As noted above, central governments tend to use their media to cling on to political power. Even the so-called press comes under direct control of central authorities. A UNESCO study expresses the concern in the following words: 'Where local and regional communication systems exist in the less developed world, they come under the spell of the national system which is dominated by commercial and often powerful international and domestic political interests.'[36] This uneasy co-existence of two economically and culturally different sections of people in one and the same country could create, and has in fact created, explosive social tensions. With the traditional media systems playing the role of the intermediary between the central political powers and the rural communities, the danger can be avoided, especially if a form of dialogue between the two sections can be established.

But community media do not have to preclude all forms of modern communication systems. Specific and concrete uses of modern communication systems for rural community projects include the Zambian housing project set up in 1974, using such modern media systems as videotape, film, photographs, slides, and printed matter; the 'Tanzania Year 16' pilot project (1971–73), aimed at assessing the effects of Ujamaa, using portable video to solicit feed-back; two pilot projects in Kenya concerned with a wide spectrum of social problems – for example, environmental education, agricultural and health issues – using tape-cassette. But for traditional communication systems to function as solid alternative systems, a measure of autonomy becomes necessary. The 'folk media', to use another term for non-commercial media systems, can only function effectively on their own terms. Like the Africa guinea-fowl and the Chinese Giant Panda, these systems do not propagate themselves in the domestic confinement of modern systems. Under these conditions, they lose their cultural relevance through the use of non-indigenous languages; their power as media for the spontaneous identification of the people with folk symbols suffers as well. Their acceptability by the people and their capacity as a means of creating self-awareness and confidence through self-expression are invalidated. Furthermore, with a commercial value attached, they become expensive to implement.

It is perhaps for this reason that Evelyn Voigt and Rajive Jain caution against allowing external forces to dictate the terms of service to community media, and against the adoption of media which 'create community dependence on external technical assistance or finance'.[37] At the same time, one should be careful not completely to reject any use of modern equipment in the traditional communication systems. A better attitude is to make room for the use of simple but dispensable modern tools while at the same time developing traditional communication equipment capable of independent existence. The autonomy of traditional communication systems is important for another reason. The perceived need for modern equipment leads to confused notions about 'technology transfer', whereby Western technology is dished out to Third World countries without regard to their real development needs. Sadly the danger of this technology has not yet been acknowledged in many developing countries, where 'communication needs for authentic self-development often run counter to the images and imageries of development imposed from foreign sources and through the mass media.'[38]

The emphasis on the recourse to locally available resources as the basis of development efforts is an emphasis on self-reliance, leading to the evolution of a 'unique and practical style of mass interpersonal communication'.[39] But self-reliance, when not narrowly conceived, extends beyond material considerations. It presupposes a set of alternative methods, ideas, formulations, policy decisions, etc. The variety of alternatives entails not just development strategies but also communication methods and concepts corresponding to new approaches to social progress. What Rajni Kothari, the Indian specialist on communication and development, calls 'Communication for Alternative Development', and Magaga Alot designates as 'Alternative Journalism' are examples of what is involved.

Before we turn to that crucial issue, however, let us briefly revisit some conventional concepts of development.

Based on the so-called 'Diffusion Theory', the concept rested on four basic assumptions: economic development is achieved through industrialisation, measured by the degree of urbanisation and determined quantitatively by means of the Gross National Product (GNP); the means by which development is attained is characterised by capital-intensive technology imported from industrial countries; economic planning is the exclusive business of private financial institutions, mainly banks, stockbrokers and similar profit-making economic organisations; underdevelopment is to be perceived as an internally generated low economic activity; conversely there are no external causes to underdevelopment. Consequently, the role of communication under these assumptions was seen as facilitating the transfer of technology from development agencies to their clients in the Third World, and propagating ideas of 'modernisation' which that technology was supposed to administer. But as matters have turned out, the 'Less Developed Countries' (LDCs) have not reached the point of take-off, and the occasion is nowhere evident. Even Wilbur Schramm, one of the most ardent proponents of this scheme of economic development, has been forced to retreat and reflect: 'While the developing regions have not been growing poorer in absolute terms, they have been falling farther and farther behind the industrialised world.'[40] And underdevelopment is not confined to those countries which do not dispose of natural wealth; on the contrary, the phenomenon pervades the totality of any and all those Third World countries which have adopted the capitalist system, or have had it imposed on them. It has now dawned on Western development theorists that economic development is not tantamount to social progress; hence Everett Kleinjans' definition of development as the process by which society acquires the capability to bring about and enhance the quality of life of its people. In other words, a distinction has had to be made between the mere production of material wealth and the quality of life, such as environmental protection, health care, healthy and edible food, the quality of education, and so on. For this reason, new thinking on development became inevitable, bringing with it the concept of 'New Alternatives', for which the following considerations have been taken into account: the equitable distribution of wealth; the role of traditional social systems; an emphasis on human labour vis-à-vis labour-intensive technology; self-reliance; the decentralisa-

299

tion of development planning; popular participation in decision-making; the elimination of external causes of development.

We are now addressing the concept of 'appropriate models' of development. It should be noted from the outset that the basis of any 'appropriate model' is an act of political resolution. Information policy, for example, is by definition a political articulation of the role of communication in the economic life of a particular society. Every country defines its information policy in terms of its socio-economic needs. For example, Marc Porat defines American information policy as follows: 'Information policy attends to the issues raised by the combined effects of information technologies ... on market and non-market events.'[41] Which simply means that American information policy is based on technological change, on how this change affects the economic set-up of the country in general and what measures should be taken to regulate economic activities in the wake of this change. The ideological content of this formulation is obvious; and it is to his credit that, unlike most American writers, Porat allows for a different definition of information policy: 'We would err grievously to suggest that all information policy has technological roots ... In fact some of the most insightful debate about the future course of our society occurs at the ideological level.'[42]

There may not be a ready-made alternative theory of development, as Rajni Kothari argues, but the ideological dimension of information policy assigns politics the task of creating appropriate conditions under which such a theory can be formulated. It is true, moreover, in deference to Kothari's observation, that the political process does not take place in an atmosphere of rationality, that there are forces beyond the control of the Third World, which determine the course of events in the actual world. But, in order to assist the process of articulating alternative perceptions of development, the first step to take in that direction is the identification of the obstacles militating against the effort. Kothari is of the opinion that the compartmentalisation of politics and economics is responsible for the failure of the solutions attempted on development problems:

> We cannot separate the dimensions of conflict over resources, the militarisation of the world, the global economic crisis, or human rights and the survival of cultures. We have to see them as inter-related. This has been another fault of development thinking; to see problems and solutions in unidimensional terms. It is why the solutions have not worked.[43]

If the separation of politics from economic issues is incompatible with the development goals of the Third World, this has been necessitated by divorcing popular participation from the process of development as a whole. This is a logical consequence of a conception of national development which relegates the majority of the people to secondary activities, leaving decision-making to a bureaucratic machinery which is not in touch with the real problems. An alternative approach must make popular participation indispensable, integral to the entire process of development. And this means that participation by the whole community must cover the analytical as well as the conceptual stages, not just the actual implementation of projects. This point deserves special emphasis for obvious reasons:

> Since development processes affect the life of the widest sections of the population, their participation in the elaboration or discussion of such aims seems expedient. To recruit their involvement, it is necessary to inform properly not only the experts but also wide segments of population of developing countries on development issues.[44]

That is the first point to make about alternative approaches to development. The second is that the concept of development should not be confined to individual projects, for example, agriculture, health, family planning, etc. The process has to encompass the totality of social transformation. It should be a process consciously designed for building a new society. This might mean formulations of development policies aimed at dispensing with foreign aid altogether. Indeed Kothari considers foreign aid an ineffective method of solving development problems.[45] This clears the way for self-reliance as the main alternative to foreign aid. It is interesting to note in this regard that the idea has been subjected to a lot of ridicule in reference to the Tanzanian Ujamaa. No argument is proffered here to suggest that the Tanzanian experiment was a total success. What is clear, though, is that the piquancy with which Western analysts and their faithful African protégés have danced to the tune of the 'failure of Ujamaa' betrays ideological hostility. But one development specialist refuses to join the chorus of condemnation. Andreas Fuglesang argues that Ujamaa was the correct development policy to adopt in search of an indigenous alternative. The so-called 'failure' cannot be seen in isolation from other factors which were not always internal in origin. At any rate the final impact of Ujamaa on the Tanzanian development efforts may still be on its way.[46]

Before we conclude this discussion, the special role of communication in development alternatives should be recapitulated. The biggest problem facing the Third World in general, and Africa in particular, is the biased and erroneous depiction of these countries by the Western media. This is to be expected since the Third World does not possess independent news agencies of its own, which is to say, information networks capable of independently collecting information, and interpreting and disseminating it themselves. The Third World sees itself and the rest of the world through foreign eyes: AFP (the French Press Agency), AP and UPI (Associated Press and United Press International – both USA), Reuters (UK) and Tass (USSR). These international news agencies are supported by two types of satellite systems, supplying information along the East–West ideological schism. There is the American international satellite system, INTELSAT, which operates such information-gathering facilities as ATS-I (the so-called Applications Technology Satellite), PEACESAT (intended for use by developing countries to facilitate ideas on development), and ATS-6 (with which India has been experimenting for its programmes on communication for national development). Then there is the INTERSPUTNIK satellite system set up in 1971 by the Soviet Union.

Information-gathering and supply is basic to social development. Africa is a classical victim of the lack of indigenous information supply. The intergovernmental conference on communication policies in Africa, a UNESCO-sponsored undertaking, produced this depressing picture: The circulation of newspapers per 1,000 inhabitants in four different regions of the world was given in 1980 as follows: Africa, 14; Asia, 64; Latin America and the Caribbean, 70; the industrialised countries, 312.[47] It may take time before African countries, or the Third World for that matter, can develop their own modern communication systems. Meanwhile, as one African communication specialist argues, African countries cannot afford to sit and wait to produce advanced communication technologies. They can and should initiate alternative means of generating information for development. They can and should avail themselves of the existing traditional communication systems. Some of the uses into which traditional forms of communication can be put include transmission of cultural activities through folk tales, dramatic arts, visual or graphic arts, songs, proverbs, riddles, dances, music, games, etc. The training of youth for development purposes and encouragement of popular participation in development issues are activities within the competence of traditional forms of communication. The use of traditional communication systems implies also development of appropriate technology. By all standards, this has to be simple *vis-à-vis* the heavy-duty type imported from abroad. As the working paper on communication policies in Africa correctly states:

A simple technology which corresponds to needs, uses the most flexible methods and equipment, can provide immediately the type of solution needed, is often a more viable investment for developing countries than advanced technology, whose cost may be exorbitant and which may turn out to be unsuitable for the aims it is supposed to serve.[48]

I have argued in this essay that communication and development theories share common concepts. Ideas on development are transmitted through media designed to carry definite ideological concepts, which makes communication technology an ideological implement. Therefore, no definition of communication, or development for that matter, can be undertaken without regard to the ideological affinity of the 'definer'. That is why Kwasi Wiredu's definition of communication as the 'transfer' of 'thought-content' is incomplete since it leaves out altogether the interpretive model adopted for the purpose. It has also been the contention here that culture constitutes the framework within which any concept of development must be viewed. This view leads to the point about indigenous perceptions of development and the attendant need for popular participation in development planning and decision-making. All this brings in politics as the key to successful formulation and articulation of an alternative theory of communication appropriate to the development needs of the Third World. In their simplicity, traditional communication systems are the best suited media for the purpose. While acknowledging the very real obstacles militating against the development of traditional communication systems, it is our belief that these can be overcome, precisely through socio-political reorientation.

References

1. K. Wiredu, 'A philosophical perspective on the concept of human communication', in *International Social Science Journal*, vol. 32, no. 2, 1980, p. 199.
2. D. B. Patel, 'Mass Communication and Development in Africa', in A. A. Mazrui, H. Patel (eds.), *Africa in World Affairs: The Next Thirty Years* (New York: The Third World Press, Joseph Okpaku Publishing Company Inc., 1973), p. 191.
3. See *inter alia* G. Gerbner, 'Communication and Social Environment', in *Communication: A Scientific American Handbook* (San Francisco: Freeman, 1972).
4. H.-P. Kraft, *Information und Kommunikation im Rahmen staatlicher Wirtschaftspolitik* (Berlin: Nicolaische Verlagsbuchhandlung, 1978), p. 13.
5. Ibid.
6. M. L. de Fleur, S. Ball-Rokeach, *Theories of Mass Communication*, 4th ed. (New York, London: Longman, 1982).
7. Ibid., p. 16.
8. Ibid., p. 145.
9. E. Kleinjans, 'Foreword', in W. Schramm, D. Lerner (eds.), *Communication and Change: The Last Ten Years – And The Next* (Honolulu: The University Press of Hawaii, 1976), p. ix.

10. W. W. Rostow, *The Stages of Economic Growth: A Non-Communist Manifesto* (Cambridge: Cambridge Univ. Press, 1960).
11. M. Bedjaoui, *Towards a New International Economic Order* (Paris, New York, London: UNESCO/Holmes & Meier Publishers, 1979), p. 67. See also, F. Perroux, *A New Concept of Development* (Paris: Croom Helm/UNESCO, 1983).
12. Bedjaoui, *Towards a New International Economic Order*, p. 67.
13. M. Marchaha, *Contribution à la Notion de Droit International Public du Développement*, unpublished thesis, University of Nice, 1976 (quoted in Bedjaoui).
14. C. Furtado, 'A Global View of Development Process', in *Different Theories and Practices of Development* (Paris: UNESCO, 1982), p. 79.
15. M. Tehranian, 'The Curse of Modernity; The Dialectics of Modernisation and Communication', in *International Social Science Journal*, vol. 32, no. 2, 1980.
16. Ibid., p. 255.
17. Ibid.
18. A. Cabral, *Unity and Struggle* (New York, London: Monthly Review Press, 1979), p. 149.
19. Y. Littunen et al., 'Cultural Problems of Direct Satellite Broadcasting', in *International Social Science Journal*, vol. 32, no. 2, 1980, p. 292.
20. D. Senghaas, 'Kultur und Entwicklung – Überlegungen zur aktuellen Entwicklungspolitischen Diskussion', in *Zeitschrift für Kulturaustausch*, no. 14, 1984, part II, p. 423.
21. A. Jamal, 'The Cultural Dimensions of Development: National Cultural Values versus Transnational Cultural Domination', in *Development Dialogue*, nn. 1–2, 1984, p. 80.
22. E. J. Hobsbawm, *The Age of Capital: 1848–1975* (New York: Charles Scribner's & Sons, 1975).
23. Ibid.
24. Ibid.
25. H. Elsenhans, 'Peripherer Staat und abhängige Entwicklung', in *Zeitschrift für Kulturaustausch*, p. 340.
26. O. Kreye, 'Neue Internationale Wirtschaftsordnung oder Rekolonialisierung der Entwicklungsländer?', in *Zeitschrift für Kulturaustausch*, pp. 327–8.
27. M. U. Porat, *Information Economy: Definition and Measurement* (Washington DC: US Department of Commerce/US Government Printing Office, 1977), p. 2.
28. T. W. Halina, 'Communication and the Economy: A North American Perspective', in *International Social Science Journal*, vol. 32, no. 2, 1980.
29. G. N. S. Raghavan, 'Do Mass Media Reach the Masses? The Indian Experience', in Z. Morsy (ed.), *Media Education* (Paris: UNESCO, 1984), p. 34.
30. G. Mytton, *Mass Communication in Africa* (London: Edward Arnold, 1983), p. 74.
31. For an extensive contemporary account, see P. M. Lewis (ed.), *Media for People in Cities: A Study of Community Media in the Urban Context* (Paris: UNESCO: COM 84/WS-7, 1984), p. 2.
32. R. Colin, 'La communication sociale et la participation populaire au développement entre tradition et modernité', in Huynh Cao Tri et al., *Stratégies du développement endogène* (Paris: UNESCO, 1984), p. iii.
33. Perroux, *New Concept of Development*, p. 172.
34. De Fleur and Ball Rokeach, *Theories of Mass Communication*, p. 234.
35. See Mytton, *Mass Communication in Africa*.
36. Lewis, *Media for People in Cities*, p. 8.
37. Ibid., p. 112.
38. Tehranian, 'The Curse of Modernity', p. 254.
39. M. Alot, *People and Communication in Kenya* (Nairobi: Kenya Literature Bureau, 1982), p. 211.

40. W. Schramm, 'An Overview of the Past Decade', in W. Schramm, D. Lerner (eds.), *Communication and Change*, p. 2.
41. Porat, *Information Economy*, p. 207.
42. Ibid., p. 210.
43. R. Kothari, 'Communications for Alternative Development: Towards a Paradigm', *Development Dialogue*, vol. 1, no. 2, 1984, p. 19.
44. V. A. Vinogradov et al., 'Towards an International Information System', in *International Social Science Journal*, vol. 33, no. 1, 1981, p. 21.
45. Kothari, 'Communications for Alternative Development', p. 18.
46. A. Fuglesang, 'The Myth of People's Ignorance', in *Development Dialogue*, vol. 1, no. 2, 1984.
47. *Intergovernmental Conference on Communication Policies in Africa: A Working Paper*, UNESCO, 1980, p. 330.
48. Ibid., p. 51.

RANGGASAMY KARTHIGESU

Television as a Tool for Nation-Building in the Third World: A Post-Colonial Pattern, Using Malaysia as a Case-Study

Television in the Third World

When we speak about television in the Third World, we speak, of course, of the television phenomena in recently independent, developing countries. Both these conditions have influences on the circumstances of the inception of the television service in a particular country, its growth pattern, its programme content, its functions in society and its character as a whole. If in the western world, the cradle of television, its complexities are still in the process of being researched, its benign and malign qualities still being determined, it can safely be said that in the developing world it still remains an enigma, sometimes misunderstood and misused. There its complexities have acquired even more dimensions, its effects even more complications than in the western world. In the developing Third World countries, television has not had a chance to evolve harmoniously in the cultural habitat of the foster countries. In many cases it is an alien form rudely thrust into the landscape, a transfer, not of technology alone, 'but also to the transfer of socio-cultural institutions with economic and political implications, institutions ready-packaged with organisational and programme formats and even contents.'[1]

It is in the developing countries that television is formally assigned with objectives and directions by authorities which are often governments. For example in Thailand those aims and objectives were described, by the government in 1965, as follows:

a) to promote national policy and common interests in the areas of politics, military affairs, economics and social welfare, b) to promote the loyalty of the citizens to the country, the religion and the King, c) to promote the unity and mutual cooperation of the army and its citizens, and d) to invite citizens to retort to and oppose the enemy, including those doctrines which are dangerous to the security of the nation.[2]

The Iranian objectives in 1966 were described as fostering integration, the stimulation of cultural development and widespread popular participation. But in 1968 they were redescribed in connection with the Fourth National Development Plan (1968–1972), reflecting a definite planning purpose:

> The first of the goals listed is 'to publicise national development affairs from their social, political and economic point of view', and 'to keep the public informed of current events'. Strengthening national unity and fostering culture, art, education and entertainment are lower on the list reflecting, as might be expected, the bias of planners.[3]

In Egypt, a 1959 presidential decree established for broadcasting the following goals, among others: a) elevating the standards of the arts; b) strengthening national feeling and social cooperation, spreading solidarity between social groups and supporting accepted tradition; c) discussing social problems and strengthening spiritual and moral values; and d) reviving the Arabic literacy, scientific and artistic heritage.[4]

The Malaysian objectives, showing a technocratic bias as in the case of Iran above, are more open about their purpose:

> a) to explain in depth and with the widest possible coverage the policies and programmes of the government in order to ensure maximum understanding by the public; b) to stimulate public interest and opinion in order to achieve changes in line with the requirement of the government; c) to assist in promoting civic consciousness and fostering the development of Malaysian arts and culture; and d) to provide suitable elements of popular education, general information and entertainment.[5]

It is in the developing countries, then, that television's potential and energy is claimed to be harnessed with specific objectives in mind. But often the overt noble objectives seem to be accompanied by a 'hidden agenda' which has more to do with keeping the national leadership stable than with the general purposes of national integration, or the exclusion of negative western influence.[6] The setting up of specific aims and objectives by these developing countries for their broadcasting services, draws attention to a paradox that goes with it. Although these countries import the bulk of technical equipment and know-how as well as programme production techniques and training methods from the west, not to mention the actual bulk of programmes for their television, the aims and objectives that they assign to their broadcasting services bear no resemblance to the way television operates in the countries from which the hardware and know-how have been imported. These aims and objectives relate to making broadcasting a propaganda instrument for government, and since governments and parties that control governments become more or less indistinguishable, it becomes a propaganda tool for parties in power as well. In style, at least, this is closer to the Soviet and Communist Eastern Europe model of broadcasting, countries with which many of the developing countries, at least many of those in the Commonwealth, may find themselves at odds ideologically, rather than the western world with which they associate themselves.

Yet it is not from the Communist World that developing countries of the Third World adapted such a style. Instead it was the British colonial masters themselves who facilitated the exclusive servicing of government in power, by setting up a 'colonial model' of broadcasting (as opposed to the 'metropolitan model' of the BBC) for the purposes of imposing their dominance and authority over their colonial subjects, and subsequently bequeathed it to the newly independent governments. True, this was before the advent of television, but in most countries television was merely an extension of radio services and the same mould was used to cast the new service.

Today almost all the countries in the developing world have accepted and acquired television. Whether governments profess to like it or not, urbanisation and modernisation have enticed them to bring television into their countries. The volume of research is growing on what television does to societies in these countries, their government, politics, culture, arts, education and life-styles, in short to the shaping and developing of their mind. But research is still inadequate to fully describe these effects. Awareness is high and research questions are beginning to be asked, but answers have only begun to emerge in skimpy patterns. True, answers are skimpy even in the western world, but there at least the volume of research is substantial.

Television in Malaysia
Malaysia, whose television service is treated as a case-study for this essay, is a typical example of a former British colony going through its economic and social development stages since its independence in 1957. It is a country with rich natural resources in rubber, tin ore, tea and timber, from which Britain and its industries benefited enormously. But in the independence movement that swept Asia, Britain had to leave. As it did so, it made certain that the power and authority was transferred to an élite rule, predisposed to be loyal and friendly to Britain, so that existing economic relationships would not be harmed.

Broadcasting (then only radio) was part of the power structure transferred to the new government and was designed to provide the same service that it provided for the colonial government, namely to safeguard and strengthen the authority of government. There was a strongly built-in partiality towards people and parties in power. There wasn't a moment, of course, when any of the involved parties thought of transforming a 'colonial service model' to an 'independence service model', or to a 'metropolitan model' like the BBC, even granting that the new government was still looking to emulate Britain, following whose example it set up the institution of the monarchy and two houses of parliament on the Westminster model. But this did not happen, and Malaysia was not unique in not doing so. India, Pakistan as well as Sri Lanka, all of which obtained their freedom earlier, did not do so either.

Television Malaysia was the creation of the independent government of Malaysia in 1963, and it was moulded in exactly the same way as Radio Malaysia. Six years of independence had not provided enough confidence or inspiration to institute a more open and liberal broadcast service. Fostering that attitude was the overhanging threat of armed Communist insurgence in the country which, although quelled earlier, had not been completely eliminated. Television grew rapidly in the country, covering the urban centres and spreading to rural areas. Sets were affordable by a substantial number of rural people and viewership rapidly grew. Today it is the foremost medium of entertainment and information in the country with 93% of the adult population of the country able to receive it.[7]

The television phenomenon in Malaysia, as elsewhere in the developing Third World countries, is a complex one. Its inception, growth and development are full of implications for the politics, governance and social development of Malaysia as a whole. Though it is not untypical of the pattern of growth and development of television generally in the developing countries, some unique points need to be noted. In a finely balanced multiethnic society composed of Malays, Chinese and Indians – who in turn make up a multireligious society following Islam, Buddhism, Hinduism and polytheism – all of which have to exist in an uneasy and imposed harmony, the role and function of the television medium becomes more complex indeed. Even at a superficial level it is not difficult to see that many more pressures are at play on the operation of the medium than is often the case, and that unique responses are required to produce some kind of order and pattern.

The external pressures that influence television in developing countries, such as the market pressure that supplies and controls equipment and technical know-how as well as programme packages are recognised easily – although not yet in their full impact. Internal pressures such as recreational demands created by urbanisation and modernisation, ideological demands created by political struggles, and informational demands created by the new thrust to integrate and develop the countries have also been recognised although, again, not in their full impact. Less recognised and studied are the implications it holds for multiracial living, religious harmony and the economic and cultural aspirations of underprivileged and minority groups within the country.

All those implications relate to Television Malaysia's presumed contribution, as a communications institution, to nation-building – the transformation of Malaysia from a colonial country into an independent, sovereign state. That television, along with other mass media, has a vital role to play in nation-building has been asserted in recent forums, the main ones being the UNESCO-sponsored conferences in Asia and Africa. We can now examine some of the concepts on nation-building that have arisen from those forums.

Nation-Building Concepts and the Role of Communications Institutions
It can be said that mass communication can be used as an important tool in the Third World to help newly independent countries attain nationhood. However, this is independent of two important factors: who owns and controls the channels of mass communication, and what their concepts are towards nation-building. It is on those concepts and objectives that the potential contribution of the media rests. In most Third World countries, the owners and controllers of the channels of communication happen to be governments and it is their concept of nation-building that determines the behaviour of the media.

Let us examine, first of all, some of the more restricted concepts of nation-building as perceived principally by governments. In a meeting of UNESCO experts on the dynamics of nation-building in Malaysia in 1979, participants provided examples of what government understanding might be with regard to nation-building.[8] Some governments, for example, conceptualise the problem of nation-building principally in terms of state-formation rather than national integration – in terms more of a creation of a bureaucracy, the military, or a technocracy – rather than in terms of a conscious effort to cultivate civil liberties through popular participation in various levels of the emerging polity. Alternatively, the exact opposite may take place: we may have a situation in which those who run the government become conscious about the need for national integration to a point where they assume that if they place race and ethnic relations on the agenda of their concerns, of if they undertake research on them, the problem will invite easier resolution.

Both cases have implications for the media. A rigid attitude to existing cultural and ethnic diversity, as in the first example, may give rise to much unreflective communication which may fan primordial sentiments. In the second case, the media may become the conduit for high-visibility national integration campaigns launched by the authorities to regulate race and ethnic relations, rather than to reconstitute them in such a manner that such communal affinities do not overshadow the daily living. The restrictive concept of nation-building – that of state-formation – often gets railroaded by governments in the name of national unity, which is considered a precondition to national economic development, creating in its path cultural and social casualties from the minority groups.

Other concepts of nation-building are more broad and far sighted and they are held principally by social critics. For instance, nation-building may be seen as involving a redefinition of sub-national identities and the emergence of a larger unity. In the process smaller identities could be preserved. But it is necessary to establish their linkages with society at large, to emphasise their interdependence. Here, the purpose of communication is to raise consciousness, achieve mobilisation, and ensure participation of the masses as a way of building a nation of liberated people controlling their own destiny.

Others view the term 'nation-building' as referring not merely to national integration, but also, and more importantly, to a vision of a truly autonomous and democratic society. Nation-building, they believe, should not mean just a working-out of an arrangement of accommodation and co-existence among ethnic groups – it should involve a welding together of a human community based on universal human values.

A UNESCO-sponsored intergovernmental conference on communication policies, hosted by Malaysia in Kuala Lumpur in 1979, produced the 'Kuala Lumpur Declaration' which affirmed the role of communication in nation-building in the following terms:

> The mass media of the developing countries bear a responsibility for contributing to the common task of nation-building and to the further development of the cultural identity of peoples and ethnic minorities, so ensuring national cohesion and creating abilities to derive the utmost benefit from enriching influences coming from outside.

Elsewhere the document also declared:

> We therefore call for participation of people and individuals in the communication process and for more freedom and autonomy for and the assumption of greater social responsibility by mass in-

311

formation media and at the same time for greater individual responsibility by and protection of those who run the media and prepare messages for circulation.[9]

The Kuala Lumpur Declaration, and the additional recommendations, for which Malaysia naturally was a signatory as a member country and participant, urged a constructive liberalism to which the country was actually becoming more distant, at least as far as its treatment of television policies was concerned, as we shall see later in the essay. For now it is sufficient to say that there are numerous instances in the history of conceptualisation, inception and development of the organisation of TV Malaysia which would point to the government actually moving in the opposite direction to the spirit of the Kuala Lumpur Declaration and the UNESCO recommendations.

Television Malaysia as a Government Department
When a government department, serving, supposedly, a neutral role in carrying out government policies made out by its political masters, is also expected to play a media role, especially an adversary role as characterised by western media concepts, the conflicts that arise are serious ones indeed. But these conflicts, to be properly understood and appreciated, are best studied by employing the concepts of organisational theories in public administration. Organisational theories suggest that the civil service is 'neutral'. But this does not mean that it is powerless. Writing about government departments in Britain, Pitt and Smith suggest that although the constitutional role assigned by the laws and conventions to civil servants should be non-political, officials in government departments serve their ministers with loyalty and neutrality (regardless of the minister's party) and proffer advice from positions of authority based on professional knowledge and experience.[10] They thus help in giving advice on public policy options and do play a role in policy-making.

But in a developing country where there is no convention of vigorous profferring of advice from the civil service to ministers – particularly to those from ruling parties enjoying absolute majority in parliament and, as in the case of Malaysia, monopoly of government since independence – the civil service may become docile and subservient; indeed, it is more likely that it becomes over-zealous in carrying out the orders of its masters.

Admittedly there have been very few studies on broadcasting institutions that investigate them from the government department angle. Such studies as exist seem to suggest that operating the broadcasting medium as a government department prevents it from achieving its full potential as a contributor to national development. Masani's broad survey of broadcasting in India suggests that government control over broadcasting is indeed restrictive and, in spite of sincerity in declared development goals, fails to achieve them satisfactorily. She found that caution and conformity were encouraged, rather than initiative and creativity, and that political control over such institutions may attain authoritarian proportions.[11] Studying Nigerian broadcasting (although from a somewhat different perspective to this paper) Ugboajah found that it expressed and responded to government economic philosophies well, rather better than the Nigerian private press. But its commitment to government-oriented development inhibited it, unlike the press, from vigorously discussing matters of social controversy.[12]

Studies relating to Malaysian broadcasting are rare. Those studying its dynamics as an institution are rarer still. Grenfell studied the audiences available for all mass media in Malaysia and suggested that such a study was a starting point for attempts to evaluate mass media roles in developing countries. Leslie Sargent, studying communications structure in Malaysia, provided a useful survey of the state of the media. In his review of a number of studies done at Universiti Sains Malaysia he found that 30–50% of readers/listeners discount government information as being unreliable, and that this may be an inevitable feature of a system in which government itself controls the communication media.[13] While these studies look at what is happening at the receiver end, they hardly shed any light on what happens at the production end. Lowe, in two enlightening studies, provided material closer to the theme of this thesis.[14]

In the former he studied the policy environment that prevailed in implementing programme strategies in the Development and Agricultural Section and found:

In the implementation of these policies and goals, Radio Television Malaysia's response has been mainly administrative, resulting in a plethora of controls, censorship, script vetting and incessant previews of pilots and actual programmes. Production staff were issued with directives of do's and dont's. In the ensuing years since 1969, the collective wisdom of the department's production staff has consisted of an informal inventory of decision rules. These have no official or written existence and they are not known as 'rules'. They are rules that we have inferred from the remembered collection of examples of prohibitions arising from the department's self-censorship exercises. Together these form the sum total of the organisation's knowledge of what is acceptable for transmission.

Lowe drew the conclusion that there existed a gap between the promise and fulfilment of Radio and Television Malaysia in the de-

313

velopment and agricultural services. He showed that the actual pace of the country's development outstripped the facilities and quality of the broadcasting organisation; that the organisational rigidities constrained its performance; and that communication planners and development planners failed to integrate their activities.

In the latter study the same theme was expanded to cover the entire decision-making process of the management and the role played by political, religious, racial and even feudal influences on them. The study concluded:

> Most of the decisions on local media content came from sources outside of the professional structure of the department. The main influence of these decisions are therefore external and they emanate from the delicate political position of the country resulting in delicate racial balances as shown by compromises and different emphasis on questions of religion, language and culture.

Nation-Building and Newly Independent Nations
The newly independent nations of the Third World can pursue their nation-building objectives in several ways. These mainly depend on how they perceive their priorities at any given moment in their history. For most countries 'development' means primarily economic development, that is increase in agricultural and manufacturing output, raising the levels of income, improving infrastructure, etc. As these are seen as a prerequisite for better living, they are pursued with vigour by building a strong bureaucracy, increasing the professionalisation of its members and creating a technocracy. There is massive intervention by governments in planning the economy with the help of technocrats and seeking to direct the private sector towards set goals.

When such policies are being pursued, governments seek to persuade the communication institutions to go along with them, not only asking the media not to create questions and controversies so as not to distract from government objectives, but also actively to support them. These policies have found support among social scientists. Pye, touching on the role of communication policies in development programmes in countries which were in different stages of growth, argued that they needed somewhat different policies:

> Some countries are at the stage of very early transition, and all instruments of communication must be directed to giving support to the legitimacy of the government and to the administrative structures of states which will in time be necessary for guiding further social changes. In these countries with a small modernised elite the weight of communication policies should be on the side of

314

protecting the freedom of these leaders and strengthening their influences throughout society.

... At this phase the communications media must be used to reinforce an essentially tutelary process; the time is ripe for broad appeals and for the repetition of theses basic to civic education.

Pye qualified this support, of course, with the hope and understanding that at later stages adjustments would be made, so that different emerging interests would be accommodated and the wider population would be provided with opportunities to express their views to the élite: 'Clearly the communication policies appropriate at one stage of development may be either irrelevant or, more likely, directly damaging if applied at other stages.'[15]

An inherent danger in communication media providing unquestioning support and strength to governments in the early period of national development, or state-formation, is that such an attitude might be frozen and may not evolve into the later stages of liberal and more horizontal communication between the rulers and the ruled, as is hoped by the social critics. As Lerner's 'illuminating' example showed, governments, when they become frustrated by the task of converting an inert or unwilling people into their way of thinking, may conclude as Major Salem of Egypt did: 'Personally I am convinced that the public was wrong.'[16]

Taking the particular case of broadcasting, which tends to be appropriated by governments in the name of using it for national integration and national development, the initial attitude of providing the government with unquestioning support becomes normative and persists even into an advanced stage in democracy as was evidenced in many Third World countries including India, Pakistan and Sri Lanka. Their ownership tends to remain a firm monopoly of the government.

When this happens, especially when it is carried over from a colonial administration without a break, it tends to have a sense of loyalty of slavish proportions. Altaf Gauhar, reviewing the decay of political institutions in Pakistan, writes that the civil and armed services developed two strong traditions from the British – a strong sense of duty and an irresistible will to obey government orders.[17] These traditions were meticulously preserved by governments of many newly independent nations to their own advantage. These were the traditions employed to run the civil service and, *ipso facto*, the broadcasting services. In the long run, such a tradition is injurious and constricts the growth of a democratic society.

It should also be remembered that many of the privately owned newspapers played an important part in the struggle for independence while broadcasting was still serving the colonial masters. Ironically both seem to play the same roles even after the attainment of independence, only this time against and for an indigenous ruling group. In an independent country this state of affairs invites comparison between the two forms of the media, causing considerable loss of credibility and respect for the broadcast media in the public mind.

Frank Campbell, speaking from his own working experience in the Guyanese government, saw full justification in the utilisation of nationally owned media for the support of the government's initiated development plans. In 1974, the Guyanese Prime Minister (later President), Forbes Burnham, said that 'the government has a right to own sections of the media and the government has the right, as a final arbiter of things national, to formulate a policy so the media can play a much more important part than it has played in the past in mobilising the people of the country for the development of the country'.[18]

Campbell, agreeing with the sentiments of the Prime Minister, however, saw that in approaching these policies of development, the Guyana government applied a 'pathologically narrow interpretation and implementation' to development journalism. He added:

By a narrow implementation of government-supporting, development-oriented media policy, I refer, *inter alia*, to the withholding or distortion of the truth; an apparent fear that the criticism of inefficiency in even a single governmental agency would cause the agency, if not the entire government, to crumble; and an exclusion of the opposition voice from the media, especially the print media.[19]

Thus we find that the confidence to ease controls so that the media, especially those owned by governments, can display a more liberal attitude, particularly to include criticism of government, does not come easily to Third World governments.

There is no doubt that a tight grip on the media may produce short-term benefits. It was pointed out at the UNESCO experts meeting on dynamics of nation-building in Kuala Lumpur, referred to earlier, that when economic crises grip the 'new states', the ruling elite may fall back upon communal loyalties to remain in power despite the failure of their leadership and the challenge to their legitimacy and may exploit media towards that end.

Under the ownership and control of governments, media can also be easily exploited during political crises to support and bolster ruling powers. This was the case during the Malaysian emergency following the 1969 race riots. The findings of research into the dynamics of nation-building in 1980, following the UNESCO experts' meeting referred to earlier, identified several noteworthy examples. They included the constitutional crisis in Fiji in 1977, the Panchayat or multi-party system question in Nepal in 1980, the Philippines situation in the 70s and the October revolts of Thailand in 1973 and 1976.[20]

Television Malaysia and Nation-Building

From its inception in 1963, TV Malaysia was constituted as a fully-fledged government department. A committee of government officials appointed by the Prime Minister in 1960 recommended the introduction of the television service along the same lines as the radio service. No other body outside the government was consulted. Parliament was not asked for authority or advice in establishing the service, and the scheme was simply announced as a *fait accompli*.

The government engaged consultants from Canada under the Colombo Plan and they offered advice only on the technical and practical matters in implementing the service and did not concern themselves with policy questions to do with control of the medium and its operations. They assumed, perhaps correctly, that it was not their place to question the wisdom of the proposed system. But it was worth observing that they cooperated to bring about a system that was monopolistic, completely government controlled, and with limited or no access for views outside government approval, although they themselves came from a system that would not have tolerated such a state of affairs.[21]

The reason for their failure to do so is perhaps found in the attitude then prevalent among some development scientists that a developing country in its early stages needed the support of its media and that the freedom of the leaders needed protection and their influence in society needed strengthening. Even if it were so, defining the television service as a full government department was tantamount to appropriating the principal mass medium of the country for the exclusive service of a small segment of a democratic society. So firm was this definition that liberalisation even at a gradual pace was rendered impossible. Political control was firmly entrenched from the beginning. Throughout its life TV Malaysia has served and propagated the policies of the ruling government, never allowing any other view to be expressed in its broadcasts.

As civil servants TV Malaysia staff live in fear of offending their own bosses as well as any section of the audience that the government wish not to offend. Such attitudes reduces all programme material to the least objectionable level, thus making them conformist, reverential to tradition and authority. One example is the production of a Malay drama, entitled *Awang Ku Sayang* (*Awang, My Dear*) in

1965, which was found to be offensive to Malay university students and was quickly amended on the instructions of the minister.

That a bureaucratic system does not encourage people with initiative was also demonstrated in 1977 in the case of the public affairs programme *Panorama*, looking at inefficiencies in government departments, which was taken off the air abruptly. Although the programme was 'well balanced', with responses by officials to criticisms aimed at them by the public, government felt too sensitive at the unfamiliar washing of their dirty linen in a government medium. This abrupt act of censorship of a programme which had won wide public support and praise from the print media must have had a significant impact on young and idealistic producers in the organisation.

During every general election Television Malaysia has been fully deployed for the purposes of helping to retain the incumbent government in power. Protests in parliament by opposition members were simply shrugged off. But in 1969 the practice was so blatant as to be counter-productive. A government review committee itself was compelled to point out that:

The programmes offered by TV Malaysia during the last two weeks of the election campaign, for example, had given an over-exposure of very direct propaganda ... thus giving rise to adverse reactions from the urban people....

But, ironically, the conclusions of the same committee were that 'the government media ... had not been utilized to the fullest possible extent with regard to the propagation of the masses, of government policies, aims and objectives, and the rationale behind such matters'. Its recommendations were that of tightening up controls, increasing the role of decision-making at the Ministry level and thus lessening the powers of heads of departments.[22]

A consequence of such a government stranglehold on the medium was an erosion – in fact, an almost total obliteration – of the trust that the intelligentsia could place in the organisation. A survey of critical opinion among a section of critical observers of Television Malaysia showed that when questioned about the way in which TV Malaysia was serving the interests of the various levels of society in Malaysia, the vast majority of the respondents felt that it was not doing so with fairness. Respondents were asked if TV Malaysia served the interests of the various races, economic classes and the political groups fairly. Over 95% answered 'no' to all three questions. As far as race was concerned the majority felt that it served only the Malays; in terms of economic class, it served only the middle and upper income groups; as far as political groups went, it served only the ruling party.

Gaining access to TV Malaysia to air views different from those of the government was also felt to be impossible. Only 27% of all respondents felt that such an access was available to them while 60% felt they could not get access. Even such limited access was restricted to messages such as appeals for charities and publicity for a gathering or meeting of groups of a non-political kind. Of the people surveyed, 96% expressed dissatisfaction with the ownership and control of TV Malaysia, stating, among other reasons, that it was bad for the growth of democracy, hid the truth from the people, and discouraged free thinking. When offered a choice of systems, nobody preferred a wholly government owned and operated system, the majority (60%) preferring a government channel along with private channels.

Throughout its history, TV Malaysia has been able to produce only about half of its output of programmes on its own, depending on imported programmes for the rest. In fact the proportion of imported programmes has been, except for the initial few weeks of its existence, consistently higher. An analysis of the programme content for a regular week at five-year intervals shows the following figures:

	% of local programmes	% of imported programmes
1963 (first week of January)	48.8	48.1
1969 ,,	37.9	55.0
1974 ,,	42.1	45.6
1979 ,,	38.4	49.1
1984 ,,	41.9	48.2

(The rest, making up the 100% in each analysis, consisted of station breaks, advertising and other announcements)

The bulk of foreign programmes came from the United States and other western countries such as Great Britain and Canada. A small proportion came from Australia, Japan and Middle Eastern countries. All these imported programmes could be classified as entertainment except those from the Middle East which were principally chosen for their Islamic religious content.

The presence of a large number of foreign programmes in Malaysian television broadcasts was not something that the organisation was happy about. Although it had been helpful in selling TV sets and drawing larger audiences, it attracted criticisms from the élite and placed TV executives in a position where they had continually to express their dislike for foreign programmes, but had to defend and justify them nevertheless.

Soon after television was launched in Malaysia, there was a demand to expand the hours of broadcast as well as the area of transmission coverage. Filling those hours with meaningful and interesting local programmes was beyond the capacity of the production staff, whose numbers, resources and training could not match the demand. On the other hand, imported programmes from the west were cheap, readily available and attractive. It was as if the TV set were created for the mass-produced, syndicated films market, and the audience – particularly the urban ones, the largest groups to possess the sets – seemed to hunger for it more and more. In short, popularising the TV service depended upon using more and more imported programmes, and that was the course taken by the organisation.

But the choice was made carefully. Only those programmes that were deemed to be entertaining were chosen. Among documentaries only those of a pragmatically educational nature were chosen. Ideological programmes of all kinds were kept away. In any case the preference was for the bulk of entertainment programmes from Hollywood – the detective thrillers, cartoons and soap operas – for these were the time-fillers. But the organisation was not unconscious of the values with which even the entertainment programmes were laden. These programmes were censored and frequently edited for scenes of a sexual nature, including kissing, unruly or violent behaviour, etc. Many programmes were rejected outright, too. On the whole they were sanitised as much as possible before they were allowed on the screen.

But it was not possible to keep these influences away altogether. Western concepts of sex and physical intimacy had become so liberalised and pervaded entire programmes that it was impossible to exclude them entirely. These concepts remained intertwined with dialogue and action essential to the story line so that editing them out would have made the programmes incomprehensible.

Thus, in spite of criticism by the élite and promises by government ministers that foreign programmes would be reduced to less than 40% of airtime, which has remained a target on paper throughout TV Malaysia's history, foreign programmes continued to dominate the airtime of TV Malaysia.

In this regard, even the government of the most authoritarian period of the 1969 emergency proved to be helpless. The Minister during this period, Ghazalie Shafie, admitted to parliament that there was an excess of foreign films – 'many of them tasteless' – on TV Malaysia, and that he had contractual obligations to show them. In a political and ideological context, local civic and religious organisations attempted to get TV Malaysia to ban American programmes in protest against America's support for Zionism. Although the government was officially in sympathy with the cause, again TV

Malaysia was not able to ban or even reduce the number of programmes being shown, citing 'contractual obligations'. On the other hand,
if government ever showed signs of reducing the foreign, particularly
American, imports, urban recreational interests spoke up vehement-
ly, and were supported by the free press, as evidenced during the
period of 1976–79. Thus the hold of the foreign imported programmes proved to be far stronger than government pretended it to be,
and government remained quite helpless in reducing it.

The effects on the audience remain uncertain. In a survey of critical opinion, 71% of the respondents expressed satisfaction with the
foreign entertainment shown on TV Malaysia while 29% were dissatisfied. But the satisfaction rate did not seem to have anything to
do with the values intrinsic in these programmes. It seemed to stem
from the belief that the audiences were discriminating enough, that
its grounding in its own culture and education would prevent it from
being influenced, and that the audience would accept the programmes as no more than entertainment. If they learned anything at all,
it would be the good things ostensibly intended in the programmes.
Nevertheless, some critics maintained that there was an emphasis in
these programmes on consumerism, materialism, violence and sex,
the sheer viewing of which in itself was a pollution of the mind of the
audience.

More significantly, the organisation was criticised for its emphasis
on selecting these programmes. The fact that the bulk of the selection was entertainment programmes rather than documentaries of
an educational kind showed that the organisation was aiming for
popularity rather than edification. In this sense it can be argued that
television was being used to create a narcotic effect on the population
instead of being used to raise its level of thinking and intelligence.

If the organisation seemed quite helpless in controlling the
amount and content of foreign imported programmes, and partially
resigned to whatever effect they might have on the audience, this
was not the case with local programmes, whose direction and contents were entirely under its control. At its inception and during the
formative years 1963 to 1969, TV Malaysia did not have any specific
declared objectives, other than the broad and universal ones of information, education and entertainment. However, all these three
aspects were carefully vetted and directed to ensure that they conform to the government's wishes. During those years the television
service was regarded almost casually. Tunku Abdul Rahman, the
former Prime Minister and founder of television in Malaysia, viewed
it principally as an entertainment medium and wished to see more
sports – including horse racing – included. In local programming
Bahasa Malaysia, the national language, was given a prominent
place, but Chinese and Tamil programmes were interspersed in the

same channel, generally creating a feeling of acceptance (although vernacular audiences remained dissatisfied).

However, as an after-effect of the 1969 racial disturbances, at which the government felt a loss of confidence and authority, the government moved to tighten all controls on national affairs. A committee was set up to reorganise and 'strengthen' the Ministry of Information and Broadcasting.

The government moved to tighten up all bureaucratic controls on the organisation and established specific objectives for the department, the two most important of which were to explain in depth and with the widest possible coverage the policies and programmes of the government in order to ensure maximum understanding by the public, and to stimulate public interest and opinion in order to achieve changes in line with the requirement of government. Every aspect of local programming was made to serve those objectives. News became a list of ministerial pronouncements and achievements. Opposition views were totally blacked out. Current affairs programmes were placed under a newly created Public Affairs Division whose head was a politician close to the Malay party, UMNO (United Malay National Organisation) and whose responsibility was directly to the minister, and overrode the powers of the director-general.

In the face of criticisms of all these, the attitude of the government hardened considerably. A minister, answering questions in parliament said:

> Allow me to remind the honourable member that radio and television are government instruments, and whether other parties can or cannot participate in discussion programmes depends on the government in considering the situation and subject under discussion.[23]

Local news and current affairs programmes thus became politicised and remained biased not only in favour of government, but, by a process of extension, in favour of the ruling party. Again, although the ruling party was an alliance of three ethnic parties (later expanded to embrace a number of other parties as Barisan Nasional), the medium served only the centre of that alliance, namely UMNO. Even government ministers of the component parties, if they held views other than UMNO on national affairs, were not allowed to express them on television. Thus it was obvious that TV Malaysia had become the voice of a particular segment of Malaysian society, the dominant and ruling elitist one, to the exclusion of all others.

To this end all vestiges of ethnic minority programmes of Chinese and Indian languages were eliminated from channel one, which was made the prime channel, and were confined to channel two. Here

322

their position was frozen and their growth regressed. The following table shows locally produced Chinese and Tamil programmes as a percentage of all locally produced programmes:

	Chinese	Tamil
1963 (initial weeks)	14.8	3.0
1974	4.1	4.0
1979	2.5	2.2
1984	2.4	2.2

It is clear that there was intentional neglect of ethnic minority programmes in the way they were allowed to languish in spite of protests from the respective communities. It was significant that even this limited amount of programming in vernacular languages consisted of news and current affairs programmes, serving as government propaganda, and not cultural programmes. Ethnic cultural programmes, in the form of singing and dancing by local talent, constituted a very insignificant portion of predominantly Malay programmes. Thus, clearly, if Chinese and Tamil languages were retained at all in TV Malaysia, it was only to use them for government propaganda purposes, not to promote any form of ethnic culture or language *per se*.

TV Malaysia also took an astringent view of religion other than Islam. All vestiges of non-Islamic religions were removed while Islam was promoted with vigour. Islam was also seen as a pervasive element in programming and all programmes came to be closely scrutinised to see that they did not deviate from Islamic beliefs. Non-Islamic religious elements, such as the showing of idols, other symbols and forms of worship, were censored. In local dramas Islamic ideology became a central theme and a controlling element.

It is apparent that such programming, in fact serving only sections of society although it was called national, was divisive and created aversion among a substantial section of the multiracial, multilingual and multireligious audience. It was not surprising that in our survey of critical opinion some expressed the view that in spite of all the shortcomings attributed to imported foreign programmes, they were the ones that united the TV audience, while local programmes, in effect, divided them into segments.

Almost all aspects of local programming proved to be highly unsatisfactory to the respondents in our survey of critical opinion on TV Malaysia (see Table I).

Some of the consequences of this kind of restrictive and manipulative programming were that the government achievements were glorified and failures whitewashed, there was no room for opposition to argue out its point of view, and no opportunity for citizens to make

TABLE I

*TV Malaysia opinion survey: Percentage of respondents satisfied and
dissatisfied with local programmes by programme category*

Programme category	% satisfied	% dissatisfied
Local entertainment	25	75
Local news	15	85
Current affairs	0	100
Topical matters in current affairs:		
Ruling political parties	35	65
Opposition political parties	0	100
Government affairs	50	50
Human rights and civil liberties	0	100
Minority races	4	96
Women's rights	21	79
Youth	50	50
Trade Unions	0	100
Rural Development	54	46
Consumer rights	10	90
Business and industry	4	96
Arts and culture	25	75
Religion	25	75

their own judgment on, or participate in discussions of, national
issues. The atmosphere for democracy to grow and mature was rendered consequently unconducive.

Implications for Nation-Building of Government Control on Programming Policies

At the beginning of this essay we examined some of the prevalent
concepts of nation-building, both the restricted view of state-
formation as held by many governments of newly independent nations, with their attempts to build and strengthen bureaucratic institutions, and the broader view of building of democratic societies,
as held by social critics, preserving diversity and encouraging a larger unity with universal human values.

The priority for newly emergent governments, at least those modelled on the parliament and parties of Westminster, remain economic
and social development as seen through the ruling party's eyes and
the eyes of the voting majority that put the party in power. They
develop institutions as infrastructure to galvanise the party in power. Social justice, beyond party and party supporters' interests, is a
largesse they cannot afford. At best they can only justify what they
do for party and electoral benefits as part of a larger scheme of social
justice which they envisage, but which will take time to see im-

plemented. But time for those governments is available only in five-year or four-year chunks, i.e. between general elections; hardly enough to plan and execute a satisfactory order of social justice. Their plans therefore frequently do not progress beyond mere rhetoric. All they aim to do is to satisfy a large enough segment of society that will vote them in at the next election. It may involve a measure of social justice, but this is neither imperative nor a high priority.

One may be given to wondering, given such a high degree of dissatisfaction among its intelligentsia, what it is that such mass media organisations are aiming to achieve on behalf of government, their masters. We can only hypothesise that by providing voluminous and repetitive propagandistic information, the organisations wish to convince the bulk of their audience – ignoring the disgruntled elite minority – of the government's choices of actions in national affairs.

What these organisations contribute to nation-building, then, is the strengthening of the ruling powers, their ideologies and their protagonists, thus encouraging the growth of a rigid and authoritarian style of bureaucracy, promoting segments of the nation's social composition at the expense of the growth of other segments. The ignored segments of society belong not only to minority races, but also to minority religions and language groups, as well as the disadvantaged minority classes.

References

1. E. Katz, G. Wedell, *Broadcasting in the Third World* (London: Macmillan, 1978), p. v.
2. F. Wheen, *Television: A History* (London: Century Publishing, 1985), p. 43.
3. Katz and Wedell, p. 28.
4. D. A. Boyd, *Broadcasting in the Arab World* (Philadelphia: Temple Univ. Press, 1982), p. 22.
5. Radio Television Malaysia, *Radio Television Malaysia Handbook* (Kuala Lumpur, 1975).
6. See J. Lent, 'ASEAN Mass Communications and Cultural Submission', *Media, Culture and Society*, vol. 4, 1982, p. 4.
7. Malaysian Ministry of Information, *Communications Systems in Malaysia* (Kuala Lumpur, 1983).
8. UNESCO/Asian Institute for Broadcasting Development, *Report on Experts Meeting on Dynamics of Nation-Building* (Kuala Lumpur, 1979).
9. UNESCO, *Final Report on Intergovernmental Conference on Communication Policies in Asia and Oceania* (Kuala Lumpur, 1979).
10. D. C. Pitt, B. C. Smith, *Government Departments: An Organisational Perspective* (London: Routledge and Kegan Paul, 1981).
11. M. Masani, *Broadcasting and the People* (New Delhi: National Book Trust, 1976).
12. F. O. Ugboajah, 'Nigerian Mass Media in Social Crisis', in F. O. Ugboajah (ed.), *Mass Communications, Culture and Society in West Africa* (London: Hans Zell, 1985).
13. L. Sargent, *Communication Structure in Malaysia* (Penang: UNESCO/Universiti Sains Malaysia, 1976).

14. V. Lowe, *Development Broadcasting: Policy Implementation and Programming in the Development and Agricultural Section of Radio and TV Malaysia* (Singapore: AMIC, 1980); V. Lowe, J. Kamin, *TV Programme Management in a Plural Society* (Singapore: AMIC, 1982).
15. L. W. Pye, 'Communication Policies in Development Programmes', in L. W. Pye (ed.), *Communication and Political Development* (Princeton, NJ: Princeton Univ. Press, 1963).
16. Quoted in D. Lerner, 'Towards a Communication Theory of Modernisation', in Pye (ed.), *Communication and Political Development*.
17. A Gauhar, 'Pakistan: A Case Study in Institutional Decay', in P. Lyon, J. Manor (eds.), *Transfer and Transformation: Political Institutions in the New Commonwealth* (Leicester: Leicester University Press, 1983).
18. F. Campbell, 'The Practical Reality of "Development Communication"', in *Intermedia*, vol. 12, 1984, p. 26.
19. Ibid., p. 26.
20. See UNESCO Regional Office for Education in Asia and Pacific, *Dynamics of Nation-Building*, vol. 2: *Communication in Crisis Situations* (Bangkok, 1985).
21. See E. S. Hallman, *Broadcasting in Canada* (London: Routledge and Kegan Paul, 1977).
22. Ministry of Information and Broadcasting, Malaysia, *Report of a Special Committee Appointed by the National Operations Council to Consider the Reorganisation and Strengthening of Ministry of Information and Broadcasting* (Kuala Lumpur, 1969).
23. Parliament Report, 18 March 1971.

Contributors to this Volume

DR RIVA S. BACHRACH
Kibbutz Child and Family Clinic, Tel-Aviv, Israel.

ROSALIND BRUNT
Lecturer in Mass Communications; Director of Studies, MA in Women's Studies; and Director, Centre for Popular Culture, Sheffield City Polytechnic, England. Publications include *Silver Linings* (co-editor; Lawrence and Wishart, 1981).

DR BARRIE GUNTER
Head of Research, Research Department, Independent Broadcasting Authority, London, England. Extensive publications on a variety of mass communications research topics include *Poor Reception: Forgetting and Misunderstanding Broadcast News* (Erlbaum, 1987).

MARY BETH HARALOVICH
Assistant Professor in Media Arts Department of The University of Arizona. Her 1984 doctoral dissertation was entitled *Motion Picture Advertising: Industrial and Social Forces and Effects, 1930–1948*. She has published extensively in the fields of film and television.

DR L. ROWELL HUESMANN
Professor of Psychology and Chairperson, Department of Psychology, University of Illinois, Chicago, USA. Executive Secretary, International Society for Research on Aggression. Consulting Editor, *Journal of Applied Psycholinguistics*. Extensive publications include *Growing Up to be Violent: A Longitudinal Study of the Development of Aggression* (co-author; Pergamon, 1977) and *Television and the Aggressive Child: A Cross-National Comparison* (co-author; Erlbaum, 1986).

MARTIN JORDIN
Lecturer in Film Theory and Sociology of Literature, Sheffield City Polytechnic.

RANGGASAMY KARTHIGESU
Lecturer and Chairman of Film and Broadcasting Programme, Universiti Sains Malaysia, Penang, Malaysia. Professional experience in Malaysian radio and TV 1961–1975. Currently completing doctoral research at the University of Leicester Centre for Mass Communication Research, on *Two Decades of Growth and Development of TV Malaysia and an Assessment of Its Role in Nation-Building*.

SONIA M. LIVINGSTONE
Lecturer in Sociology of Media at Brunel University. Forthcoming publications include *Audiences and Interpretations* (Pergamon Press).

DR PAULA W. MATABANE
Graduate Associate Professor, Department of Radio, TV and Film, School of Communications, Howard University, Washington DC, USA. Papers and articles on audiences, ethnicity and learning from media. 1985 doctoral dissertation on *The Relationship between Experience, Television Viewing and Selected Perceptions of Social Reality: A Study of the Black Audience.*

KYALO MATIVO
Lecturer, School of Journalism, University of Nairobi, Kenya. Current doctoral research on African communications, particularly traditional communication systems in Kenya, at University of California, Los Angeles, USA. Wide range of critical and creative writing.

WILLIAM H. MELODY
Director, Economic and Social Research Council Programme on Information and Communication Technologies, London, England; on leave from the interdisciplinary Faculty of Applied Science, Simon Fraser University, Vancouver, Canada. Formerly Chief Economist, US Federal Communications Commission, 1966–71, involved in pioneering research into the relationship between telecommunications and computers. Professor, Annenberg School of Communications, University of Pennsylvania, 1971–76. His extensive publications include *Children's Television: The Economics of Exploitation* (Yale, 1973) and *Information and Communication Technologies: Social Science Research and Training* (ESRC, 1986).

DR ERIC MICHAELS
Lecturer in Humanities, Griffith University, Queensland, Australia. Formerly Research Fellow, Australian Institute of Aboriginal Studies, Canberra, where he conducted *The Remote Television Project*, a four-year field study of the introduction of television to a traditionally-oriented Aboriginal community. Publications include *The Aboriginal Invention of Television* (Canberra: AIAS, 1986) and *For A Cultural Future: Francis Jupurrurla Makes Television at Yuendumu* (Sydney: Artspace, 1987).

GUILLERMO OROZCO-GOMEZ
Doctoral candidate in education, Harvard University, Cambridge, USA; visiting researcher, Universidad Autonoma Metropolitana, Mexico City. Publications include *La asignacion de recursos economicos en la educacion en Mexico: un proceso technico en un contexto politico* (co-author; SEP, F. Barros Sierra, 1983).

PATRICIA M. PALMER
Head of Research and Development for ABC Television, Australia. Former teacher and lecturer. Publications include *The Lively Audience* (Sydney: Allen & Unwin, 1986) and *Television and Teenage Girls* (New South Wales Ministry of Education, 1986).

DR CLAUS-DIETER RATH
Collaborator, Institute for Semiotics and Communication Theory, Free University, Berlin, West Germany. Publications on television and the anthropology of everyday life include *Rituale der Medienkommunikation* (co-editor; Guttandin and Hoppe, 1983). Contributor, *Television in Transition* (BFI, 1986).

PADDY SCANNELL
Senior Lecturer, School of Communication, Polytechnic of Central London, England. Has published widely on the history of broadcasting. Editorial Board member of *Media, Culture and Society*.

KIM C. SCHRØDER
Assistant Professor, Department of English, University of Roskilde, Denmark. Co-editor, *Mediekultur*, journal of the Danish Mass Communication Research Association. Publications include *The Language of Advertising* (co-author; Blackwell, 1985).

IAN TAYLOR
Professor of Sociology at Carleton University, Ottawa, Canada. He has recently taught at La Trobe University, Melbourne, Australia, and been Visiting Fellow at the University of Cambridge. Extensive publications include *The New Criminology* (co-author with Paul Walton and Jock Young; RKP, 1973), *Law and Order: Arguments for Socialism* (Macmillan, 1981) and *Crime, Capitalism and Community* (Toronto: Butterworths, 1983). He is currently researching forms of socialist-feminist cultural studies in the 1920s.

DR HANS VERSTRAETEN
Research Assistant, Department of Communication Science, Free University, Brussels, Belgium. Has published widely on the political economy of the media. Publications include *Pers en macht: een dossier over de geschreven pers in België* (Kritax, 1980).

DR JACOB WAKSHLAG
Director of Primary Research, CBS Inc., New York, USA. Formerly in Department of Telecommunications, Indiana University, Bloomington, Indiana, USA, and Visiting Research Fellow, IBA, London.

Index

331